# SpringerBriefs in Computer Science

*Series Editors*

Stan Zdonik
Peng Ning
Shashi Shekhar
Jonathan Katz
Xindong Wu
Lakhmi C. Jain
David Padua
Xuemin Shen
Borko Furht
VS Subrahmanian
Martial Hebert
Katsushi Ikeuchi
Bruno Siciliano



For further volumes:
http://www.springer.com/series/10028

Markus Bestehorn

# Querying Moving Objects Detected by Sensor Networks

Springer

Markus Bestehorn
Research and Development Department
Landis+Gyr
Undermülistrasse 28
Fehraltorf, Switzerland

ISSN 2191-5768 ISSN 2191-5776 (electronic)
ISBN 978-1-4614-4926-3 ISBN 978-1-4614-4927-0 (eBook)
DOI 10.1007/978-1-4614-4927-0
Springer New York Heidelberg Dordrecht London

Library of Congress Control Number: 2012946736



Printed on acid-free paper

Springer is part of Springer Science+Business Media (www.springer.com)

*To Maike, Kurt, Tina, Max and Daniela.*

# Preface

Declarative query interfaces to Sensor Networks (SN) have become a commodity. These interfaces allow access to SN deployed for collecting data using relational queries. However, SN are not confined to data collection, but may track object movement, e.g., wildlife observation or traffic monitoring. While relational approaches are well suited for data collection, research on Moving Object Databases (MOD) has shown that relational operators are unsuitable to express information needs on object movement, i.e., spatio-temporal queries. In this paper, we study declarative access to SN that track moving objects. The properties of SN prevent a straightforward application of MOD, e.g., node failures, limited detection ranges and accuracy which vary over time etc. Furthermore, point sets used to model MOD-entities like regions assume the availability of very accurate knowledge regarding the spatial extend of these entities. As we show, assuming such knowledge is unrealistic for most SN. This paper is the first that defines a complete set of spatio-temporal operators for SN while taking into account their properties. Based on these operators, we systematically investigate how to derive query results from object detections by SN. Finally, we show how process spatio-temporal queries in SN efficiently, i.e., reduce the communication between nodes. Our evaluation shows that our measures reduce communication by 45%-89%.

# Acknowledgments

Stephan Kessler and Thomas Thieringer significantly contributed to the work presented here by implementing and evaluating our spatio-temporal query processor for Sun SPOT sensor nodes. Olga Ulmer implemented the centralized strategy for processing spatio-temporal predicated in sensor networks.

As with any success in life, one's family and friends are those who make the success possible. In particular, I would like to thank my parents, *Maike* and *Kurt*, for pushing me to stay at the university and complete a doctoral degree. In cooperation with my sister *Tina* and my brother *Max*, they always helped me to get my head up again when work frustrated me.

# Contents

# Querying Moving Objects Detected by Sensor Networks

## 1 Introduction

Many sensor-network installations (SN) observe moving objects. For instance, scientists observe animal movement [14, 37, 43], or authorities monitor soldiers, pedestrians or vehicles [24, 34, 35]. In such applications, users are interested in object movements, i.e., the queries have *spatio-temporal semantics*.

A promising way to access SN are declarative queries [9, 10, 23]. But research has focused on *relational* queries so far. Formulating spatio-temporal information needs with relational operators results in very complex query statements [25, 54]. *Moving object databases* (MOD) have solved this problem by proposing operators with concise spatial and spatio-temporal semantics.

There are several characteristics of SN that are in the way of a straightforward application of MOD concepts to SN: MOD tend to assume that information on objects and regions is complete and accurate. Data collected with SN in turn typically does not have this characteristic. First, unobserved areas due to failed nodes and the inaccuracy of detection mechanisms result in inaccurate/incomplete information on the movement of an object. For instance, laser scanners detect the distance of an object such as a vehicle to the node equipped with the scanner, but not the exact position of the object. Other mechanisms are even less accurate, e.g., acoustic vehicle detection only detects if a vehicle is in the vicinity of the node [16]. Second, MOD model regions as point sets which implies that precise information on the spatial extend of the region is available at any time. As we show, acquiring such information for many SN deployments is unfeasible or even impossible. To circumvent this problem, these SN typically observe object movement in relation to a set of nodes instead of a set of points. We refer to such a set of nodes as *zone* to distinguish it from the term *region* which denotes a point set. Since zones are a peculiarity of SN, they have not been addressed by research on MOD. Third, the inaccuracy of object detection sometimes prevents the SN from determining whether an object is inside, on the border or outside

M. Bestehorn, *Querying Moving Objects Detected by Sensor Networks*,
SpringerBriefs in Computer Science, DOI 10.1007/978-1-4614-4927-0_1,

of a region. It is challenging to provide spatio-temporal operators for SN with clear semantics for regions and zones while coping with the intricacies of object detection.

In this paper we propose *Moving Objects Sensor Databases (MOSD)*, i.e., declarative access to sensor networks that track moving objects. More specifically, we make the following contributions:

**Applicability:** Different detection mechanisms use different hardware with different properties and varying accuracy. Furthermore, deployments of SN themselves vary regarding several characteristics. We define meaningful abstractions applicable to all kinds of detection mechanisms and deployment types without sacrificing conciseness and expressiveness.

**Semantics:** We provide a set of spatio-temporal operators for SN with concise semantics. These operators allow users to express spatio-temporal queries in SN. The systematic translation of object detections into results for queries interested in object movement in relation to a zone or region is the core contribution of this paper.

**Optimality:** In some cases, the SN is unable to determine whether the movement of such an object conforms to a query or not due to the inaccuracy of detection mechanisms. We identify these cases and provide an approximate query result by dividing objects into three sets: The first set contains objects that definitely conform and the second those that definitely do not conform to the query. The third set consists of objects where the SN cannot provide a definite result. We prove that our approximation is optimal, i.e., the aforementioned translations minimize the third set.

**Efficiency:** Processing spatio-temporal queries must be energy-efficient, because sensor nodes are typically battery-powered [1, 41]. We provide two different execution strategies to compute spatio-temporal query results in-network and reduce communication by exploiting spatial correlation of object detections. Our evaluation shows that these strategies reduce communication by 45%-89% compared to processing the query at the base station.

Finding a solution to the problem addressed by each contribution is challenging itself. However, it is important to note that these underlying problems cannot be solved independently from each other one by one. This paper provides an integral approach that addresses all of them.

## 2 Applications for MOSD

We now describe two applications of object-tracking SN and provide examples for spatio-temporal queries. The scenarios illustrate the core differences between the two main classes of spatio-temporal queries in SN and introduce two important subclasses for each class.

## 2.1 Application Example 1: Surveillance

Figure 1 illustrates an application from vehicle detection and classification called "A line in the Sand" [5]. Sensor nodes track vehicles moving in an area. An example of a spatio-temporal query is "Which vehicles $\mathbf{V}_i$ have entered the restricted access region **R**?".

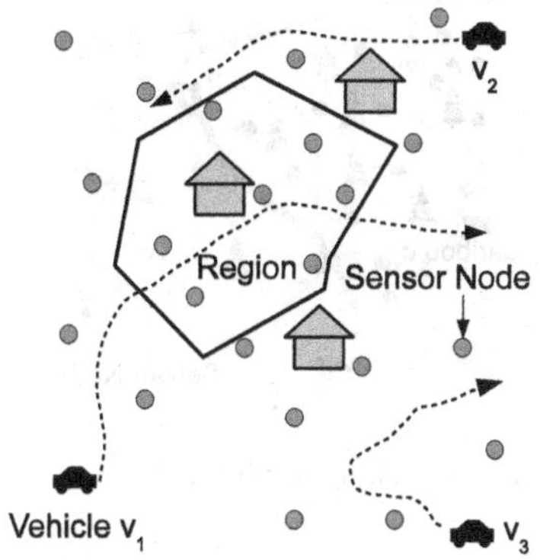


**Fig. 1** Illustration of a surveillance application

As we show in Section 3.1, there exist various mechanisms that allow the detection of objects such as vehicles, humans or animals. While some of them, e.g., radar [17], allow precise localization of objects detected, most of them only determine if an object is in the vicinity of a sensor node, e.g., microphones [11, 16]. Hence, sensor nodes might be unable to determine if an object detected is inside the region, on the border or outside. Another issue, which is discussed in [5] as well, is the possibly *uncontrolled deployment* of sensor nodes for surveillance applications: For military deployments in particular, it is often infeasible to deploy nodes manually, e.g., because the area of interest is controlled by enemy forces. Hence, sensor nodes may be dropped out of an airplane. This may result in unobserved areas [3]. Summing up, MOSD must cope with inaccurate and incomplete information on the movement of objects.

For the query above, the region **R** is a set of points that does not change over time. We call such a region *static*. Another way to define a region is by means of constraints referring to values which change over time. For example, a user could define a region as all points of space with a temperature below $0°C$. In this case, the region changes over time (*dynamic region*).

## 2.2 Application Example 2: Animal Tracking

Tracking animals at large temporal and spatial scale is important to understand their behavior [14, 33]. SN can be deployed over large areas and allow the monitoring of animals such as caribous [44, 47] without much intrusion. The following is an example of a spatio-temporal query scientists could issue: "Which caribous $\mathbf{C}_i$ have moved into the tree-covered swamp area on the south-western side of the river?"

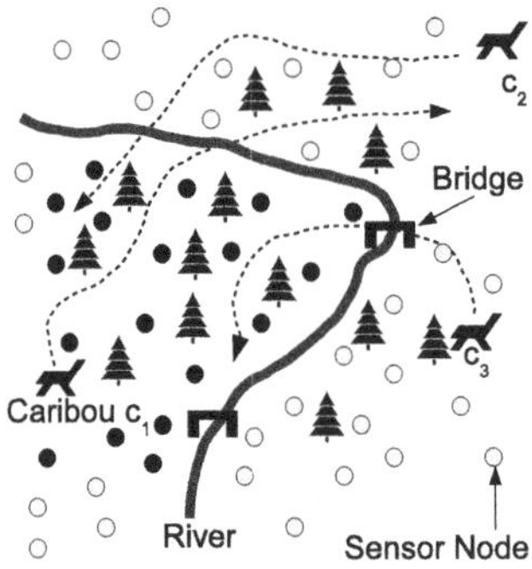


**Fig. 2** Illustration of an animal-tracking application

It is possible, but impractical, to model this swamp area as a point set. This is because such a model would require exact recording of the locations of all trees, the swamp and the river. Typically, scientists solve this problem by carefully planning the node positions and placing them manually [22]. This *controlled deployment* allows recording properties of the surroundings of each node during deployment, i.e., before the nodes start sensing. Based on this information, one can derive a set of nodes inside the area of interest, e.g., all nodes in the tree-covered swamp area on the south-western side of the river (black colored circles in Figure 2). It is sufficiently accurate for its purpose if the SN observes caribou movement in relation to this set of nodes. As stated in the introduction, we refer to such a set of nodes as *zone* to distinguish it from the term *region* which describes a point set. In Figure 2, the zone is the set of black circles. Analogously to regions, there are *static* and *dynamic zones.*

## *2.3 Scope and Assumptions*

We are interested in a declarative interface for sensor networks that observe moving objects and its efficient implementation. We study queries on the spatio-temporal relationship of a moving object and a region or zone which may be static or dynamic.

**Definition 1 (Spatio-Temporal Query):** A *spatio-temporal query for* SN is a tuple $Q = \{\mathbb{O}, \mathsf{C}, \mathbb{P}\}$:

1. **Object Description** $\mathbb{O}$: A description of objects whose movement is queried. The description must allow sensor nodes to identify relevant objects using their sensing hardware.
2. **Query Context** $\mathsf{C}$: This is a region or zone.
3. **Predicates** $\mathbb{P}$: A set of predicates and operators that define movement the user is interested in.

An object matching the description $\mathbb{O}$ is part of the result if it has moved as described by $\mathbb{P}$ in relation to the region or zone described by $\mathsf{C}$. □

Sections 3.2 and 5 will elaborate on query contexts for regions and zones respectively. The spatio-temporal predicates and operators which describe the movement of interest will be addressed in Sections 6 and 7. Note that the query definition deliberately excludes queries interested in the topological relationship of two regions, two zones, lines and regions etc., since such queries are outside of the scope of this paper.

Additionally, there are some assumptions resulting from the applications envisioned in a natural way: Nodes are stationary, i.e., they do not move once they have started sensing. Nodes are able to distinguish between query-relevant objects and irrelevant ones. This means that if the query is interested in vehicles, the detection mechanism can distinguish vehicles from other kinds of moving objects, e.g., pedestrians. This is realistic, because detection mechanisms typically are designed for a specific type of object. For example, mechanisms for the detection of animals, e.g., acoustic animal recognition [37], filter irrelevant events. Other mechanisms for animals use collars [38, 44] attached to individuals of the species observed, i.e., animals without a collar remain undetected.

In addition, the various detection technologies typically allow the identification of individuals. This is important for spatio-temporal queries. In particular, if node $\mathcal{S}_i$ detects an object, and another node $\mathcal{S}_j$ detects the same object later on, the SN can derive that the object is the same. Such an identification is typically available, e.g., through identification numbers on the collars, characteristic noise patters or ferro-magnetic signatures (see [5] for examples).

# 3 Background

This section reviews related work and introduces concepts/mechanisms our work is based on. There are three areas of research related to ours; the numbers are in line with the ones of the corresponding subsections:

3.1 Detection Mechanisms: There exist detection mechanisms for various kinds of objects. We review some of them and summarize their properties.

3.2 Moving Object Databases: MOD facilitate the processing of queries with spatio-temporal semantics. We introduce core concepts of MOD and discuss why these are not readily applicable to SN. For further details on MOD see [19, 25, 26].

3.3 Declarative Query Processing in SN: Research has shown that accessing SN declaratively is advantageous. We discuss the advantages and show why existing work is insufficient for SN that track moving objects.

Section 3.3 reviews our own previous work on spatio-temporal query processing in SN.

## 3.1 Detection Mechanisms

Object detection has received a lot of attention from research [5, 11, 17, 27, 28, 36, 37, 48, 49, 57]. For example, magnetometers have been used to detect and identify the magnetic field generated by moving vehicles [28]. Most of the research in the area aims at increasing the accuracy of detection or at efficiency, particularly if readings from several nodes must be combined to detect an object. Spatio-temporal query processing as proposed in this paper is on top of these approaches: The existing mechanisms try to detect objects. We propose operators to let users access this information declaratively. We use some of the mechanisms just mentioned for illustration.

In [37], microphones have been installed on sensor nodes to detect, classify and identify animals, in this case frogs. Similarly, one can generate sound signatures from the noise of engines and propulsion gear of vehicles using microphones [11, 49]. All these mechanisms cannot determine the exact position of the object detected. This is different with other mechanisms that allow distance estimation like Laser Scanners or even provide precise locations of objects detected, like radar [17].

[5] investigates limitations regarding detection using magnetometers and micropower-impulse (MI) radar (TWR-ISM-002-I): Their magnetometers have become desensitized over time, and this effect is even stronger if the sensor was exposed to heat. While this could be fixed by circuitry that recalibrated the magnetometers at certain intervals, the area observed by a sensor node has become significantly smaller temporarily. Furthermore, the MI-radar and the magnetometer have influenced each other when both were used simultaneously. While the documentation of the TWR-ISM-002-I [2] states a maximum range of 60 feet, the actual range has been significantly lower during their experiments. External influences, e.g., rain, reduced the range even more. Hence, one has to take into account that detection ranges change over time. This may result in areas that are temporarily or permanently unobserved even if the SN has been deployed manually.

## 3.2 Moving Object Databases

Moving object databases are based on *point-set topology* [21]. According to it, a space is composed of infinitely many points, e.g., the d-dimensional

Euclidean space $\mathbb{E}^d$. We will use $\mathbb{E}^2$ for illustrations. All concepts, those of MOD as well as our own, can be extended to other spaces or more dimensions.

Point-set topology distinguishes subsets of space, i.e., sets of points, which are called entities. There are three different types of entities: *objects*[1], *lines* and *regions*. We leave aside lines in the following, since we are interested in queries related to object movement in relation to a region/zone.

**Definition 2 (Object):** An *object* **O** is an entity that is represented by its position $\mathsf{p} \in \mathbb{E}^d$ at a given time $t$. □

A region is a point set where every point $\mathsf{p}$ satisfies a set of conditions that describe an entity covering more than one point of space, e.g., a security area or storm. We denote the set of conditions that define a region **R** as $\mathsf{C}_\mathbf{R}$ and the function that checks for a point $\mathsf{p}$ if it fulfills $\mathsf{C}_\mathbf{R}$ as $\mathsf{C}_\mathbf{R}(\mathsf{p})$:

$$\mathsf{C}_\mathbf{R}(\mathsf{p}) = \begin{cases} \mathcal{T} & \text{iff } \mathsf{p} \text{ fulfills } \mathsf{C}_\mathbf{R} \\ \mathcal{F} & \text{Otherwise} \end{cases} \tag{1}$$

Defining regions as arbitrary point sets is problematic, because such point sets could contain anomalies like dangling lines, cuts and punctures. To avoid this, [51] introduce regularization which adds or removes points from regions until the aforementioned anomalies are corrected. To ease our presentation, we assume that one condition in $\mathsf{C}_\mathbf{R}$ corrects these anomalies, i.e., all regions are assumed to be regular in the following.

**Definition 3 (Region):** A *region* **R** is a set of points which satisfy a set of conditions $\mathsf{C}_\mathbf{R}$:

$$\mathbf{R} = \{\mathsf{p} \in \mathbb{E}^d \mid \mathsf{C}_\mathbf{R}(\mathsf{p}) = \mathcal{T}\} \tag{2}$$

□

Every entity **e** partitions the space into three pair-wise disjoint subsets: the *interior* $\mathbf{e}^I$, the *border* $\mathbf{e}^B$ and the *exterior* $\mathbf{e}^E$. For a region **R**, the border $\mathbf{R}^B$ is the line that encompasses the interior $\mathbf{R}^I$. Any point of space that is neither in $\mathbf{R}^B$ nor $\mathbf{R}^I$ is part of the exterior $\mathbf{R}^E$. In the context of an object **O** positioned at a point $\mathsf{p} \in \mathbb{E}^d$, the interior $\mathbf{O}^I$ contains only $\mathsf{p}$. The border $\mathbf{O}^B$ of **O** is empty and the exterior $\mathbf{O}^E$ contains every point of space except $\mathsf{p}$, i.e., $\mathbf{O}^E = \mathbb{E}^d \setminus \{\mathsf{p}\}$. See [18, 21] for formal definitions of these space partitions.

### 3.2.1 Spatio-Temporal Predicates

The 9-intersection model [18] describes the topological relationship of two entities **A** and **B**: As illustrated in Figure 3, there are nine possible intersections of the exterior, the border and the interior of **A** with the exterior, the border and the interior of **B**, respectively. Each of these intersections is

[1] Entities represented by a single point in space are typically called *point* by publications on this subject. We refer to such an entity as *object* to clearly distinguish it from a point which is an element of space.

either empty or not. Hence, a matrix of nine boolean values identifies the relationship of **A** and **B**.

$$\begin{pmatrix} \mathbf{A}^B \cap \mathbf{B}^B \neq \varnothing & \mathbf{A}^B \cap \mathbf{B}^I \neq \varnothing & \mathbf{A}^B \cap \mathbf{B}^E \neq \varnothing \\ \mathbf{A}^I \cap \mathbf{B}^B \neq \varnothing & \mathbf{A}^I \cap \mathbf{B}^I \neq \varnothing & \mathbf{A}^I \cap \mathbf{B}^E \neq \varnothing \\ \mathbf{A}^E \cap \mathbf{B}^B \neq \varnothing & \mathbf{A}^E \cap \mathbf{B}^I \neq \varnothing & \mathbf{A}^E \cap \mathbf{B}^E \neq \varnothing \end{pmatrix}$$

**Fig. 3** 9-Intersection Model for two entities **A** and **B**

While there exist $2^9 = 512$ unique intersection matrices, only three matrices describe a possible topological relationship between an object and a region [19]. Every matrix that describes a possible topological relationship is associated with a predicate, i.e., there are three predicates that describe the relationship between an object **O** and a region **R**: Inside $(\mathbf{O}, \mathbf{R})$, Meet $(\mathbf{O}, \mathbf{R})$ and Disjoint $(\mathbf{O}, \mathbf{R})$. Figure 4 shows the intersection matrices associated with these predicates and Example 1 explains them.

$$\begin{pmatrix} \mathcal{F} & \mathcal{F} & \mathcal{F} \\ \mathcal{F} & \mathcal{F} & T \\ T & T & T \end{pmatrix} \qquad \begin{pmatrix} \mathcal{F} & \mathcal{F} & \mathcal{F} \\ \mathcal{F} & T & \mathcal{F} \\ T & T & T \end{pmatrix} \qquad \begin{pmatrix} \mathcal{F} & \mathcal{F} & \mathcal{F} \\ T & \mathcal{F} & \mathcal{F} \\ T & T & T \end{pmatrix}$$

Disjoint $(\mathbf{O}_1, \mathbf{R})$     Meet $(\mathbf{O}_2, \mathbf{R})$     Inside $(\mathbf{O}_3, \mathbf{R})$

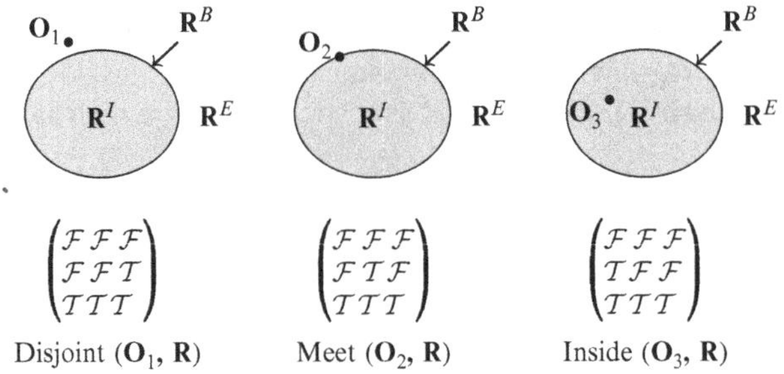

**Fig. 4** Illustrations and 9-Intersection representations of the three predicates that describe the topological relationship of an object and a region $(\mathbf{A} = \mathbf{O}_i$ and $\mathbf{B} = \mathbf{R})$

**Example 1:** The left-most matrix in Figure 4 describes Disjoint $(\mathbf{O}_1, \mathbf{R})$. As mentioned before, the border of an object is empty, i.e., $\mathbf{O}_1{}^B$ does not intersect with any partition of **R**. This is reflected by the first row of the 9-intersection matrix for Disjoint $(\mathbf{O}_1, \mathbf{R})$. The second row implies that $\mathbf{O}_1{}^I \cap \mathbf{R}^E \neq \varnothing$, i.e., $\mathbf{O}_1$ is outside of **R**. The last row of the 9-intersection matrix describing Disjoint $(\mathbf{O}_1, \mathbf{R})$ shows that $\mathbf{O}_1{}^E$ intersects with all partitions of **R**.

The matrices for Meet $(\mathbf{O}_2, \mathbf{R})$ and Inside $(\mathbf{O}_3, \mathbf{R})$ only differ from the matrix for Disjoint $(\mathbf{O}_1, \mathbf{R})$ in the second row: The topological relation of $\mathbf{O}_2$ and **R** conforms to Meet $(\mathbf{O}_2, \mathbf{R})$ if $\mathbf{O}_2{}^I \cap \mathbf{R}^B \neq \varnothing$, i.e., the object $\mathbf{O}_2$ is on the border of **R**. Similarly, $\mathbf{O}_3{}^I \cap \mathbf{R}^I \neq \varnothing$ implies that $\mathbf{O}_3$ is inside of **R**, i.e., Inside $(\mathbf{O}_3, \mathbf{R})$. ◆

### 3.2.2 Spatio-Temporal Developments

In MOD, users formulate a query by describing the movement they are interested in. To express arbitrary changes of relationships between entities, [19] defines the concatenation operator, as follows:

**Definition 4 (Concatenation):** The *concatenation* of two predicates, $\mathrm{P} \triangleright \mathrm{Q}$, is true if P is true for some time interval $[t_0; t_1[$, and Q is true at $t_1$. □

Using this operator, one can construct *sequences of spatio-temporal predicates* $\mathrm{P}_1 \triangleright \mathrm{P}_2 \triangleright \ldots \triangleright \mathrm{P}_q$. In line with [19], we refer to such a sequence as *spatio-temporal development*.

**Example 2:** In Section 2.1, the user wants to know which vehicles **V** have moved into region **R**. To fulfill the query, a vehicle **V** must be outside of **R**, then move over the border $\mathbf{R}^B$ into the interior $\mathbf{R}^I$.

$$\mathrm{Disjoint}\,(\mathbf{V},\mathbf{R}) \triangleright \mathrm{Meet}\,(\mathbf{V},\mathbf{R}) \triangleright \mathrm{Inside}\,(\mathbf{V},\mathbf{R}) \tag{3}$$

This spatio-temporal development usually is referred to as Enter (**V**, **R**).

$$\mathrm{Disjoint}\,(\mathbf{V},\mathbf{R}) \triangleright \mathrm{Meet}\,(\mathbf{V},\mathbf{R}) \triangleright \mathrm{Disjoint}\,(\mathbf{V},\mathbf{R}) \tag{4}$$

$$\mathrm{Inside}\,(\mathbf{V},\mathbf{R}) \triangleright \mathrm{Meet}\,(\mathbf{V},\mathbf{R}) \triangleright \mathrm{Disjoint}\,(\mathbf{V},\mathbf{R}) \tag{5}$$

Other sequences are constructed similarly: Equations (4) and (5) define the predicate sequences for Touch (**V**, **R**) and Leave (**V**, **R**) respectively. ◆

While infinite sequences of spatio-temporal predicates are possible, [19] has shown that it is sufficient to explicitly consider a canonical collection of 28 developments. From these 28 developments, more complex ones can be constructed by means of concatenation, as illustrated in Example 3.

**Example 3:** Suppose that a user is interested in objects **O** that enter a region **R**, move around inside the region and then leave the region. To express this using the aforementioned developments, the user concatenates Enter (**O**, **R**) and Leave (**O**, **R**):

$$\mathrm{Cross}\,(\mathbf{O},\mathbf{R}) = \mathrm{Enter}\,(\mathbf{O},\mathbf{R}) \triangleright \mathrm{Leave}\,(\mathbf{O},\mathbf{R}) \tag{6}$$

The concatenation Enter (**O**, **R**) ▷ Leave (**O**, **R**) is typically denoted as Cross (**O**, **R**). The expression in (6) translates to the predicate sequence in (7). Note that Inside (**O**, **R**) ▷ Inside (**O**, **R**) = Inside (**O**, **R**) at the junction between Enter (**O**, **R**) and Leave (**O**, **R**), since $\mathrm{P} = \mathrm{P} \triangleright \mathrm{P}$ [19]. ◆

As in [19], we provide a canonical collection of spatio-temporal developments for SN in Section 7. This allows us to limit the number of predicate sequences we must consider explicitly.

$$\underbrace{\mathrm{Disjoint}\,(\mathbf{O},\mathbf{R}) \triangleright \mathrm{Meet}\,(\mathbf{O},\mathbf{R}) \triangleright \overbrace{\mathrm{Inside}\,(\mathbf{O},\mathbf{R})}}_{\mathrm{Enter}(\mathbf{O},\mathbf{R})}\overbrace{\triangleright \mathrm{Meet}\,(\mathbf{O},\mathbf{R}) \triangleright \mathrm{Disjoint}\,(\mathbf{O},\mathbf{R})}^{\mathrm{Leave}(\mathbf{O},\mathbf{R})} \tag{7}$$

### 3.2.3 Applying Moving Object Databases to Sensor Networks

MOD model an object as a point in space. For moving objects, this implies that the position is known precisely at any point in time. Most of the detection mechanisms used in SN cannot provide this accuracy (cf. Section 3.1). There has been work aimed at processing spatio-temporal queries if object positions are only known at some instants of time [4, 13, 52, 53]. These approaches are insufficient in our context: First, they still require precise object positions from time to time. Second, they are based on relatively strict assumptions. For instance, [53] assumes that an object whose position is $p_1$ at $t_1$ and $p_2$ at $t_2$ moves between $p_1$ and $p_2$ on a straight line "at a constant speed".

To conventional notion of a border that completely encompasses a region does not readily carry over to our context. This is because (some of) the border of a region may be unobserved. For example, a user may query Enter $(\mathbf{O}, \mathbf{R})$. Let us assume that **O** moves from the outside of **R** into the region, but it is never observed on the border, e.g., because a node that has been deployed to observe the border has failed. Another problem with the border is that it is a line. The time it takes an object to move over a line is infinitely short. Capturing this moment would require an infinitely high temporal resolution of the detection hardware.

Capturing the spatial extent of regions is problematic as well in some applications. In the examples in Section 2, users formulate queries regarding the object movement relative to a set of nodes. These queries are unique to SN. Summing up, while MOD concepts serve as a foundation, significant work is required to apply them to SN.

## *3.3 Query Processing in Sensor Networks*

Research has shown that declarative access to SN is advantageous, but has been limited to relational queries so far [9, 10, 23, 39–41, 56]. For traditional database systems, research has shown that expressing spatio-temporal information needs using relational operators results in unnecessarily complex queries that are difficult to process [25, 54].

The situation is comparable for existing relational query processors for SN, e.g., TinyDB [41]. One reason is the lack of continuous or time-aware data types in purely relational systems, i.e., a value is assumed to be constant unless it is updated explicitly. For continuously moving objects, this implies frequent updates. Furthermore, relational systems lack operators and data types for point sets: Relational systems for SN only feature simple data types, e.g., integer, float or string for attributes. Storing point sets would require the decomposition of the point set into separate values stored in different tuples. Processing spatio-temporal queries would then require reconstructing these point sets prior to processing the actual query. Such a reconstruction is complex since it requires subqueries and many join operations. Summing up, storing data on moving objects detected by SN would result in frequent updates, and queries would be unnecessarily complex.

Our own work has addressed spatio-temporal queries in the context of static regions [7] and static zones [8] separately. This paper provides an integral approach that is applicable to static/dynamic regions and zones. This requires significant modifications and extensions to previous concepts. Additionally, we describe new evaluations with deployments of Sun SPOT [50] sensor nodes and optimization strategies.

# 4 Generic Model of a Sensor Network

This section provides a generic model of a SN which is fundamental for our *Applicability* contribution.

**Notation (Sensor Network):** A *sensor network* is a set $\mathsf{SN} = \{\mathcal{S}_1, \ldots, \mathcal{S}_n\}$ of sensor nodes and a base station. Every $\mathcal{S}_i \in \mathsf{SN}$ has a position $\mathsf{POS}_i \in \mathbb{E}^d$.

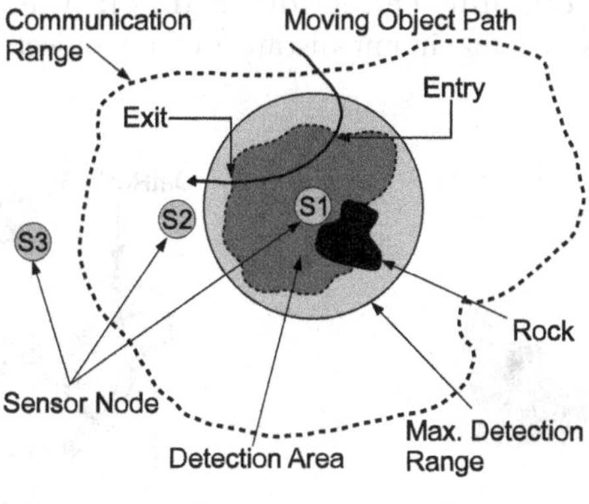


**Fig. 5** Illustration of the node model

Each node is equipped with hardware that allows it to detect and identify objects in its vicinity.

**Definition 5 (Detection Area):** The *detection area* $\mathbf{DA}_i$ of node $\mathcal{S}_i$ is the set of points $\mathbf{DA}_i \subseteq \mathbb{E}^d$ where $\mathcal{S}_i$ can detect an object. □

As discussed in Section 3.1, the detection area of a node may have any shape or size and is subject to external influences. For example[2], $\mathcal{S}_1$ in Figure 5 has been deployed close to a rock and thus cannot detect objects moving behind that rock. A *node $\mathcal{S}_i$ detects the object* $\mathbf{O}$ *at time* $t$ if $\mathbf{O} \in \mathbf{DA}_i$ at $t$.

[2] To avoid clutter in the figures, we refer to nodes in figures without subscript indices, i.e., nodes $\mathcal{S}_1, \mathcal{S}_2, \ldots$ are S1, S2, ... in the figures.

**Definition 6 (Detection Function):** The *detection function* $detect(\mathcal{S}_i, \mathbf{O}, t)$ is defined as follows:

$$detect(\mathcal{S}_i, \mathbf{O}, t) = \begin{cases} \mathcal{T} \text{ iff } \mathbf{O} \in \mathbf{DA}_i \text{ at } t \\ \mathcal{F} \text{ otherwise} \end{cases} \quad (8)$$

□

An *object* **O** *is detected at time t* if $detect(\mathcal{S}_i, \mathbf{O}, t) = \mathcal{T}$ for at least one $i \in \{1, \ldots, n\}$. Depending on the deployment, detection areas may overlap. An object within this overlap is detected by more than one node simultaneously.

**Definition 7 (Detection Set):** The *detection set* $\mathsf{DetSet}^{\mathbf{O}}_t \subseteq \mathsf{SN}$ is the set of all nodes that detect an object **O** at some time *t*.

$$\mathsf{DetSet}^{\mathbf{O}}_t = \{\mathcal{S}_i \in \mathsf{SN} \mid detect(\mathcal{S}_i, \mathbf{O}, t) = \mathcal{T}\} \quad (9)$$

□

For some detection mechanisms it is not possible to determine the detection area accurately. However, the maximum detection range is typically available prior to deployment, e.g., because the manufacturer has conducted a calibration [2].

**Definition 8 (Maximum Detection Range):** The *maximum detection range* $\mathcal{D}_{max}$ is the maximum distance of an object to a node to be detected. □

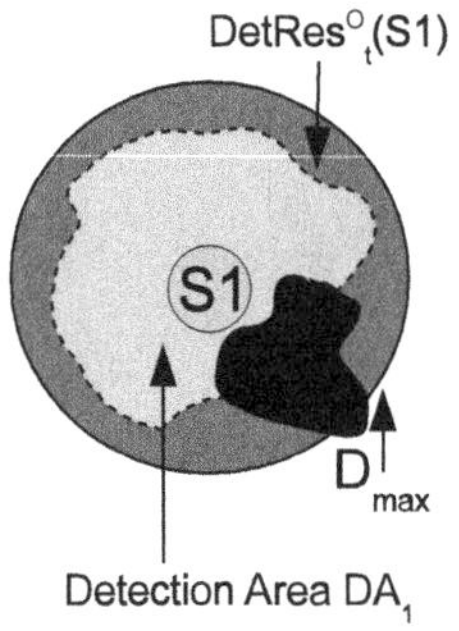


**Fig. 6** $\mathbf{DetRes}^{\mathbf{O}}_t(\mathcal{S}_1)$ based on $\mathcal{D}_{max}$

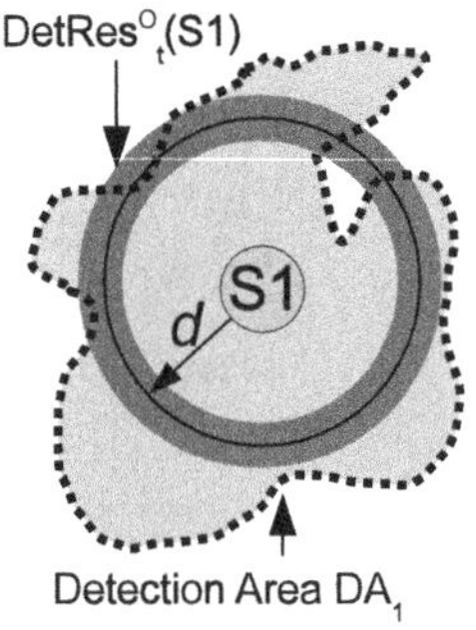


**Fig. 7** $\mathbf{DetRes}^{\mathbf{O}}_t(\mathcal{S}_1)$ with a distance estimating detection mechanism

Detection mechanisms are used to localize objects detected as accurately as possible. It depends on several factors, e.g., hardware, weather etc., how accurate such a localization is [5]. To deal with any kind of detection mechanism, we model the result of an object detection as a point set.

**Definition 9 (Detection Result):** The *detection result for an object* **O** *detected by* $\mathcal{S}_i$ *at time* $t \in \mathbb{T}$ is the set $\mathbf{DetRes}^{\mathbf{O}}_t(\mathcal{S}_i)$ of all points $\mathsf{p} \in \mathbb{E}^d$ where **O** could be according to the detection mechanism of $\mathcal{S}_i$. □

The shape and size of $\mathbf{DetRes}_t^{\mathbf{O}}(\mathcal{S}_i)$ depends on the detection mechanism, as Example 4 illustrates.

**Example 4:** Simple mechanisms like acoustic vehicle detection [11, 49] or PIR-based motion detectors cannot determine their detection area. They only determine whether an object $\mathbf{O}$ is in the vicinity, i.e., in the detection area, of a node or not. As shown in Figure 6, when $\mathcal{S}_1$ detects an object $\mathbf{O}$ at time $t$, $\mathbf{DetRes}_t^{\mathbf{O}}(\mathcal{S}_1)$ is the circle with center $\mathsf{POS}_1$ and radius $\mathcal{D}_{max}$. More sophisticated mechanisms, e.g., laser scanners, determine the distance $d$ of the node to the object. Taking into account a certain deviation $\epsilon$, $\mathbf{DetRes}_t^{\mathbf{O}}(\mathcal{S}_1)$ is ring-shaped, see Figure 7. Note that some parts of $\mathbf{DetRes}_t^{\mathbf{O}}(\mathcal{S}_1)$ in Figure 7 are not part of the detection area $\mathbf{DA}_1$ of $\mathcal{S}_1$. If $\mathcal{S}_1$ cannot determine its detection area, it cannot distinguish between points in $\mathbf{DetRes}_t^{\mathbf{O}}(\mathcal{S}_1)$ that are in its detection area and those that are not. ◆

If several nodes detect an object simultaneously, the sensor network can refine the information on the object position by intersecting the various detection results.

**Definition 10 (Possible Object Positions):** The *set of possible object positions* $\mathbf{POP}_t^{\mathbf{O}} \subseteq \mathbb{E}^d$ *of object* $\mathbf{O}$ *at time* $t \in \mathbb{T}$ is the intersection of all detection results $\mathbf{DetRes}_t^{\mathbf{O}}(\mathcal{S}_i)$ of nodes $\mathcal{S}_i \in \mathsf{DetSet}_t^{\mathbf{O}}$.

$$\mathbf{POP}_t^{\mathbf{O}} = \begin{cases} \bigcap_{\mathcal{S}_i \in \mathsf{DetSet}_t^{\mathbf{O}}} \mathbf{DetRes}_t^{\mathbf{O}}(\mathcal{S}_i) & \text{iff } \mathsf{DetSet}_t^{\mathbf{O}} \neq \varnothing \\ \varnothing & \text{iff } \mathsf{DetSet}_t^{\mathbf{O}} = \varnothing \end{cases} \tag{10}$$

□

If the detection set for an object $\mathbf{O}$ is empty, $\mathbf{O}$ is undetected. There can be various reasons for this, e.g., the object does not exist anymore or has moved into an unobserved area. Independently of the reason, the SN cannot make any statement regarding the position of the object and we model this with $\mathbf{POP}_t^{\mathbf{O}} = \varnothing$.

**Definition 11 (Communication Area):** The *communication area* $\mathbf{CA}_i \subseteq \mathbb{E}^d$ *of node* $\mathcal{S}_i$ is the set of points where a node $\mathcal{S}_j$ can receive messages sent by $\mathcal{S}_i$. □

A *node* $\mathcal{S}_i$ *can directly communicate with another node* $\mathcal{S}_j$ if $\mathsf{POS}_j \in \mathbf{CA}_i$. Communication areas can have any shape or size and may change over time. Furthermore, nodes typically cannot determine their communication area. There exist several routing protocols that determine the set of nodes that a node $\mathcal{S}_i$ can directly communicate with [20, 45]. These protocols allow forwarding of messages via multiple hops, e.g., to send results to the base station. To accomplish this, each node must store a list of nodes it can communicate directly with and some routing information about the connectedness of each neighbor to the rest of the network.

**Definition 12 (Communication Neighbors):** The *communication neighbors* $\mathsf{CN}_i$ *of a node* $\mathcal{S}_i$ are the nodes that $\mathcal{S}_i$ can directly communicate with.

□

# 5 Point Set Topology for Sensor Networks

While we borrow the concept of a region as well as its interior, border and exterior from MOD, the notion of a zone remains to be defined. We then propose a space partitioning based on zones and classify the spatio-temporal queries that occur in SN. This is a prerequisite toward the contribution *Semantics.*

A zone $\mathsf{Z}$ is a set of nodes satisfying a set of conditions $\mathsf{C_Z}$, e.g., all nodes inside a swamp area (cf. Section 2.2). Similarly to regions, we refer to the function that checks for a given node $\mathcal{S}_i$ if it satisfies $\mathsf{C_Z}$ as $\mathsf{C_Z}(\mathcal{S}_i)$:

$$\mathsf{C_Z}(\mathcal{S}_i) = \begin{cases} \mathcal{T} \text{ iff } \mathcal{S}_i \text{ satisfies } \mathsf{C_Z} \\ \mathcal{F} \text{ Otherwise} \end{cases} \tag{11}$$

**Definition 13 (Zone):** A *zone* $\mathsf{Z}$ is a set of nodes which satisfy a set of conditions $\mathsf{C_Z}$:

$$\mathsf{Z} = \{\mathcal{S}_i \in \mathsf{SN} \mid \mathsf{C_Z}(\mathcal{S}_i)\} \tag{12}$$

□

A *node* $\mathcal{S}_i$ *is inside of* $\mathsf{Z}$ if $\mathcal{S}_i \in \mathsf{Z}$, *outside* otherwise. We refer to the set of nodes that are outside of the zone as $\overline{\mathsf{Z}}$:

$$\overline{\mathsf{Z}} = \{\mathcal{S}_i \in \mathsf{SN} \mid \mathsf{C_Z}(\mathcal{S}_i) = \mathcal{F}\}$$

To define the semantics of predicates that express the topological relationship of objects and zones, it is necessary to partition the space. The core idea is as follows: Any point $\mathsf{p} \in \mathbb{E}^d$ can be either in no detection area, only in detection areas of nodes in $\mathsf{Z}$, only in those of nodes in $\overline{\mathsf{Z}}$, or in detection areas of nodes in $\mathsf{Z}$ and $\overline{\mathsf{Z}}$. Thus, every zone partitions space as follows:

**Definition 14 (Unobserved Partition):** The *unobserved partition* $\mathsf{Z}^\varnothing$ of a zone $\mathsf{Z}$ contains all points not contained in any detection area:

$$\mathsf{Z}^\varnothing = \{\mathsf{p} \in \mathbb{E}^d \mid \nexists \mathcal{S}_i \in \mathsf{SN} : \mathsf{p} \in \mathbf{DA}_i\} \tag{13}$$

□

**Definition 15 (Interior of a Zone):** The *interior* $\mathsf{Z}^I$ of a zone $\mathsf{Z}$ contains all points exclusively observed by nodes in $\mathsf{Z}$:

$$\mathsf{Z}^I = \{\mathsf{p} \in \mathbb{E}^d \mid \mathsf{p} \notin \mathsf{Z}^\varnothing \land \nexists \mathcal{S}_i \in \overline{\mathsf{Z}} : \mathsf{p} \in \mathbf{DA}_i\} \tag{14}$$

□

**Definition 16 (Exterior of a Zone):** The *exterior* $\mathsf{Z}^E$ of a zone $\mathsf{Z}$ contains all points exclusively observed by nodes in $\overline{\mathsf{Z}}$.

$$\mathsf{Z}^E = \{\mathsf{p} \in \mathbb{E}^d \mid \mathsf{p} \notin \mathsf{Z}^\varnothing \land \nexists \mathcal{S}_i \in \mathsf{Z} : \mathsf{p} \in \mathbf{DA}_i\} \tag{15}$$

□

**Definition 17 (Border of a Zone):** The *border* $\mathsf{Z}^B$ of a zone $\mathsf{Z}$ contains all points of space observed by nodes from $\mathsf{Z}$ and $\overline{\mathsf{Z}}$.

$$\mathsf{Z}^B = \{\mathsf{p} \in \mathbb{E}^d \mid \exists \mathcal{S}_i \in \mathsf{Z}, \exists \mathcal{S}_j \in \overline{\mathsf{Z}} : \mathsf{p} \in \mathbf{DA}_i \wedge \mathsf{p} \in \mathbf{DA}_j\} \tag{16}$$

□

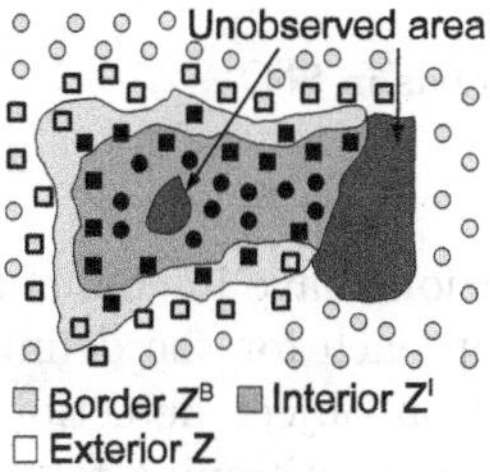


**Fig. 8** Illustration of the space partitions for a zone Z

Figure 8 illustrates this partitioning for a zone: Circles and squares[3] represent nodes. Black circles/squares represent nodes in Z while grey ones represent nodes outside of Z. Every node has a detection area of a certain shape and the space partitions of Z depend on the intersections of these detection areas.

**Lemma 1.** *The point sets* $\mathsf{Z}^{\varnothing}$, $\mathsf{Z}^I$, $\mathsf{Z}^E$ *and* $\mathsf{Z}^B$ *partition the space, i.e., every* $\mathsf{p} \in \mathbb{E}^d$ *is only in one partition.*

*Proof.* A point $\mathsf{p} \in \mathbb{E}^d$ is either included in at least one detection area or unobserved. $\mathsf{Z}^{\varnothing}$ covers all points $\mathbb{E}^d \setminus \bigcup_{1 \leq i \leq n} \mathbf{DA}_i$. The observed points $\bigcup_{1 \leq i \leq n} \mathbf{DA}_i$ are covered by one of the remaining partitions: All points exclusively observed by nodes outside of Z are covered by $\mathsf{Z}^E$. Similarly, $\mathsf{Z}^I$ covers all points solely observed by nodes in Z. All points observed by nodes inside and outside of Z are covered by $\mathsf{Z}^B$. Each of these point sets is pair-wise disjoint with the others, and thus they partition the space. ■

Lemma 1 is important: 1) It implies that there cannot exist any other other partitions. 2) The true position of an object is always in exactly one partition of a zone.

Table 1 summarizes the different types of query contexts. It contains two columns that separate the main classes of spatio-temporal queries in SN deployed to observe object movement in relation to an area of interest: The first class contains queries interested in the movement of an object in relation to a region. Queries aiming at object movement in relation to a zone constitute the second class. Both, regions and zones, can be either static or dynamic.

[3] The difference between squares and circles is irrelevant here; we explain it in Section 8.4.

| | Zone | | Region | |
|---|---|---|---|---|
| **Formula** | Node Set $\mathsf{Z} = \{\mathcal{S}_i \in \mathsf{SN} \mid \mathsf{C_Z}(\mathcal{S}_i) = \mathcal{T}\}$ | | Point set $\mathbf{R} = \left\{\mathsf{p} \in \mathbb{E}^d \mid \mathsf{C_R}(\mathsf{p}) = \mathcal{T}\right\}$ | |
| **Partitions** | $\mathsf{Z}^{\varnothing}, \mathsf{Z}^E, \mathsf{Z}^I, \mathsf{Z}^B$ | | $\mathbf{R}^E, \mathbf{R}^I, \mathbf{R}^B$ | |
| **Type** | **static** | **dynamic** | **static** | **dynamic** |
| Example | A set of unique node identifiers | Nodes measuring a temperature greater than $0°C$ | All points inside a polygon defined by GPS-coordinates | All points where the temperature is greater than $0°C$ |

**Table 1** Summary of query contexts in SN

Partitions of zones and regions have in common that they are point sets. This allows for a uniform approach for the definition of predicates and deriving result for them based on object detections as Section 6 shows. An important difference is that the partitioning for a region does not include a partition containing unobserved areas. As we show in Section 7, the lack of such a partition is the main challenge when it comes to deriving results for developments related to regions: The SN must decide if the trajectory of an object conforms to a development even if the object was undetected for some time. For example, an object conforms to Enter $(\mathbf{O}, \mathbf{R})$ (cf. Equation 3) even if the object was not detected while crossing the border of $\mathbf{R}$.

# 6 Deriving Predicate Results

In this section, we show how to derive predicate results based on object detections. By introducing *detection scenarios*, we formalize the information acquired through object detections. This constitutes our final step toward the contribution *Applicability*. The detection scenarios allow us to address the semantics of single predicates and their results, i.e., the contributions *Semantics* and *Optimality* for predicates.

## *6.1 Detection Scenarios*

When one or more nodes detect an object $\mathbf{O}$ at time $t$, the actual position of $\mathbf{O}$ is in the set of possible object positions $\mathbf{POP}_t^{\mathbf{O}}$. To derive predicate results from $\mathbf{POP}_t^{\mathbf{O}}$, one has to determine how the set of possible object positions $\mathbf{POP}_t^{\mathbf{O}}$ intersects with different partitions of the region or zone.

**Definition 18 (Detection Scenario):** A *detection scenario* $\mathsf{DS}$ is a function that returns a boolean value based on the intersection of the set of possible object positions $\mathbf{POP}_t^{\mathbf{O}}$ with the partitions of the query context, i.e., a region or zone. □

We say that a specific detection scenario $\mathsf{DS}^*$ occurs for an object $\mathbf{O}$ and a time $t$ if the detection scenario returns $\mathcal{T}$. Regardless of whether the query context is a region or zone, there are five different detection scenarios. In the following, we define the set of detection scenarios first and show that this set is exhaustive afterward.

**Definition 19** ($\mathsf{DS}^{\varnothing}$)**:** The *detection scenario* $\mathsf{DS}^{\varnothing}$ occurs if $\mathbf{POP}_t^{\mathbf{O}}$ does not intersect with the interior, exterior or border of the query context.

$$\left(\mathsf{Z}^E \cup \mathsf{Z}^B \cup \mathsf{Z}^I\right) \cap \mathbf{POP}_t^{\mathbf{O}} = \varnothing \quad \left(\mathbf{R}^E \cup \mathbf{R}^B \cup \mathbf{R}^I\right) \cap \mathbf{POP}_t^{\mathbf{O}} = \varnothing \tag{17}$$

□

**Definition 20** ($\mathsf{DS}^E$)**:** The *detection scenario* $\mathsf{DS}^E$ occurs if $\mathbf{POP}_t^{\mathbf{O}}$ is a subset of the exterior of the query context.

$$\mathbf{POP}_t^{\mathbf{O}} \subseteq \mathsf{Z}^E \qquad \mathbf{POP}_t^{\mathbf{O}} \subseteq \mathbf{R}^E \tag{18}$$

□

**Definition 21** ($\mathsf{DS}^I$)**:** The *detection scenario* $\mathsf{DS}^I$ occurs if $\mathbf{POP}_t^{\mathbf{O}}$ is a subset of the interior of the query context.

$$\mathbf{POP}_t^{\mathbf{O}} \subseteq \mathsf{Z}^I \qquad \mathbf{POP}_t^{\mathbf{O}} \subseteq \mathbf{R}^I \tag{19}$$

□

**Definition 22** ($\mathsf{DS}^B$)**:** The *detection scenario* $\mathsf{DS}^B$ occurs if $\mathbf{POP}_t^{\mathbf{O}}$ is a subset of the border of the query context.

$$\mathbf{POP}_t^{\mathbf{O}} \subseteq \mathsf{Z}^B \qquad \mathbf{POP}_t^{\mathbf{O}} \subseteq \mathbf{R}^B \tag{20}$$

□

**Definition 23** ($\mathsf{DS}^{\bullet}$)**:** The *detection scenario* $\mathsf{DS}^{\bullet}$ occurs if $\mathbf{POP}_t^{\mathbf{O}}$ intersects with two or more partitions of the query context, i.e., the detection mechanism cannot determine if $\mathbf{O}$ is inside, on the border or outside of a query context.

$$\begin{gathered}\mathbf{POP}_t^{\mathbf{O}} \cap \mathsf{Z}^E \neq \varnothing \wedge \mathbf{POP}_t^{\mathbf{O}} \cap \mathsf{Z}^B \neq \varnothing \wedge \mathbf{POP}_t^{\mathbf{O}} \cap \mathsf{Z}^I \neq \varnothing \\ \mathbf{POP}_t^{\mathbf{O}} \cap \mathbf{R}^E \neq \varnothing \wedge \mathbf{POP}_t^{\mathbf{O}} \cap \mathbf{R}^B \neq \varnothing \wedge \mathbf{POP}_t^{\mathbf{O}} \cap \mathbf{R}^I \neq \varnothing\end{gathered} \tag{21}$$

□

According to the point-set topology for regions, the border of a region is a line. $\mathsf{DS}^{\bullet}$ typically occurs in $\mathsf{SN}$ if the object detected is somewhere near the border. Only few detection mechanisms, e.g., radar, are sufficiently accurate to distinguish such an object from one on the border. Example 5 illustrates how to derive detection scenarios from object detections with a detection mechanism that cannot distinguish between objects on the border and those close to it.

**Example 5:** Let $\mathsf{SN} = \{\mathcal{S}_1, \mathcal{S}_2, \mathcal{S}_3, \mathcal{S}_4\}$, and the node positions are as illustrated in Figure 9. Each node only detects objects in its vicinity. Thus, if $\mathcal{S}_i$ detects an object $\mathbf{O}$, $\mathbf{DetRes}_t^{\mathbf{O}}(\mathcal{S}_i)$ contains all points in the circle with radius $\mathcal{D}_{max}$ and center $\mathsf{POS}_i$. Suppose each $\mathcal{S}_i$ exclusively detects a vehicle $\mathbf{V}_i$, $1 \leq i \leq 4$. Then the following scenarios occur:

$\mathbf{V}_1$ : $\mathbf{DetRes}_t^{\mathbf{V}_1}(\mathcal{S}_1)$ contains only points from $\mathbf{R}^E$. Since $\mathcal{S}_1$ is the only node that detects $\mathbf{V}_1$, $\mathbf{POP}_t^{\mathbf{V}_1} = \mathbf{DetRes}_t^{\mathbf{V}_1}(\mathcal{S}_1)$, and thus $\mathsf{DS}^E$ occurs.

$\mathbf{V}_2$ : $\mathbf{DetRes}_t^{\mathbf{V}_2}(\mathcal{S}_2)$ contains only points from $\mathbf{R}^I$. Analogously to $\mathbf{V}_1$, this means $\mathsf{DS}^I$.

$\mathbf{V}_3$ : $\mathbf{DetRes}_t^{\mathbf{V}_3}(\mathcal{S}_3)$ contains points from all three partitions of $\mathbf{R}$. This means that the detection mechanism is not sufficiently accurate to determine on which side of the border of $\mathbf{R}$ the vehicle $\mathbf{V}_3$ is. Thus, $\mathsf{DS}^\bullet$ occurs.

$\mathbf{V}_4$ : Analogously to $\mathbf{V}_3$.

Simultaneous detection of a single object can change the detection scenario. For instance, if $\mathcal{S}_4$ and $\mathcal{S}_2$ detect $\mathbf{V}_4$ at the same time, $\mathbf{POP}_t^{\mathbf{V}_4}$ is the intersection of $\mathbf{DetRes}_t^{\mathbf{V}_4}(\mathcal{S}_4)$ and $\mathbf{DetRes}_t^{\mathbf{V}_4}(\mathcal{S}_2)$. This is a subset of $\mathbf{R}^I$ and results in $\mathsf{DS}^I$.

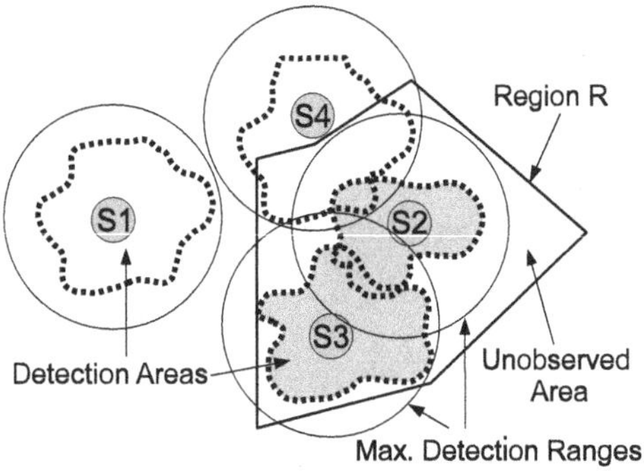


**Fig. 9** Example of detection areas, detection ranges and a region

More sophisticated detection mechanisms influence the resulting detection scenario as well. If $\mathcal{S}_3$ could determine its detection area $\mathbf{DA}_3$, $\mathbf{POP}_t^{\mathbf{V}_3}$ does not overlap with $\mathbf{R}^B$ any more. The detection scenario for $\mathbf{V}_3$ changes from $\mathsf{DS}^\bullet$ to $\mathsf{DS}^I$. ◆

The intersection of two sets $A$ and $B$ is empty, if $A = \varnothing$ or $B = \varnothing$. Thus, the detection scenario $\mathsf{DS}^\varnothing$ only occurs if $\mathbf{POP}_t^{\mathbf{O}} = \varnothing$ or if all partitions of the query context are empty.

**Lemma 2.** $\mathsf{DS}^\varnothing$ *implies that* $\mathbf{POP}_t^{\mathbf{O}} = \varnothing$.

*Proof.* The partitioning of space by regions is complete and unambiguous for regions, i.e., there always exists at least one partition that is non-empty. According to Lemma 1, the partitioning for zones is complete as well. Thus, $\mathsf{DS}^\varnothing$ implies $\mathbf{POP}_t^{\mathbf{O}} = \varnothing$. ■

**Lemma 3.** *For any object* $\mathbf{O}$ *and point of time* $t$*, exactly one of the detection scenarios* $\mathsf{DS}^{\varnothing}$, $\mathsf{DS}^{E}$, $\mathsf{DS}^{I}$, $\mathsf{DS}^{B}$ *or* $\mathsf{DS}^{\bullet}$ *holds.*

*Proof.* The lemma holds if the partitions of space, where $\mathbf{O}$ could be at $t$ based on the detection scenario currently valid, are pair-wise disjoint. If $\mathsf{DS}^{\varnothing}$ occurs, the object is undetected at time $t$. A point $\mathsf{p} \in \mathbb{E}^d$ is either in at least one detection area or unobserved. $\mathsf{DS}^{\varnothing}$ covers all points $\mathbb{E}^d \setminus \bigcup_{1 \leq i \leq n} \mathbf{DA}_i$. Thus, only those parts of space that are observed must be considered in the following, i.e., $\bigcup_{1 \leq i \leq n} \mathbf{DA}_i$. We prove the lemma for the observed part of space in the context of zones and regions separately.

In the context of a region $\mathbf{R}$, the detection scenario $\mathsf{DS}^{I}$ covers all points from $\mathbf{R}^I$. Similarly, $\mathsf{DS}^{E}$ covers all points from $\mathbf{R}^E$. $\mathsf{DS}^{B}$ occurs if the sensor network can determine that $\mathbf{O}$ is on the border for sure. Contrary to that, $\mathsf{DS}^{\bullet}$ occurs if the accuracy of the object detection is insufficient to provide a definite statement if $\mathbf{O}$ is on the border, or close to it on either side. In this case an area around $\mathbf{R}^B$ is not part of $\mathbf{R}^I$ and $\mathbf{R}^E$. All of these point sets are pair-wise disjoint.

For a zone $\mathsf{Z}$, the points covered by the respective detection scenarios are analogous to those described above. The only difference is that $\mathsf{DS}^{\bullet}$ cannot occur, because the border $\mathsf{Z}^B$ is explicitly defined as those parts of space where objects are detected by nodes in $\mathsf{Z}$ and $\overline{\mathsf{Z}}$. The lemma holds, because all parts of space are covered by the respective detection scenarios. ■

The detection scenarios abstract from the details of object detection and other issues. They also take into account simultaneous detection of an object by more than one node. The remainder of this section, we show how to derive predicate results based on detection scenarios, which is the first step towards addressing our contribution *Semantics*. Based on this, Section 7 says how to derive results for spatio-temporal developments.

## *6.2 Predicate Results for Regions*

This section shows how to evaluate predicates that describe the topological relationship of a region $\mathbf{R}$ and an object $\mathbf{O}$, given detection scenarios. $\mathsf{DS}^{E}$, $\mathsf{DS}^{B}$ and $\mathsf{DS}^{I}$ guarantee that the object detected is in a certain partition. Thus, objects detected with these detection scenarios conform to a predicate $\mathrm{P}(\mathbf{O}, \mathbf{R})$ in question or not. As illustrated in Example 5, this is not true for $\mathsf{DS}^{\bullet}$, because $\mathbf{POP}_t^{\mathbf{O}}$ overlaps with more than one partition. Objects detected according to $\mathsf{DS}^{\bullet}$ could fulfill $\mathrm{P}(\mathbf{O}, \mathbf{R})$, but this is not certain. We take this disparity regarding the certainty of object positions into account by adding a third value $\mathcal{M}$ ("maybe") to the possible results of $\mathrm{P}(\mathbf{O}, \mathbf{R})$:

- $\mathcal{T}$: $\mathrm{P}(\mathbf{O}, \mathbf{R})$ returns $\mathcal{T}$ if the SN can guarantee that $\mathbf{O}$ fulfills $\mathrm{P}(\mathbf{O}, \mathbf{R})$.
- $\mathcal{F}$: $\mathrm{P}(\mathbf{O}, \mathbf{R})$ returns $\mathcal{F}$ if the SN can guarantee that $\mathbf{O}$ does not fulfill $\mathrm{P}(\mathbf{O}, \mathbf{R})$.
- $\mathcal{M}$: $\mathrm{P}(\mathbf{O}, \mathbf{R})$ returns $\mathcal{M}$ otherwise.

**Example 6:** Continuing Example 5, suppose the user is interested in vehicles $\mathbf{V}_i$ that fulfill Inside $(\mathbf{V}_i, \mathbf{R})$. Recall that a node $\mathcal{S}_i$ can only determine if a vehicle is in its vicinity or not: $\mathbf{DetRes}_t^{\mathbf{V}_i}(\mathcal{S}_i)$ is the circle with radius $\mathcal{D}_{max}$ around the position $\mathsf{POS}_i$ of the detecting node $\mathcal{S}_i$. If node $\mathcal{S}_i$ in Figure 9 detects $\mathbf{V}_i$, $1 \leq i \leq 4$, the results are as follows:

$\mathbf{V}_1$: The distance between $\mathcal{S}_1$ and $\mathbf{R}$ is greater than $\mathcal{D}_{max}$. Thus, it is certain that $\mathbf{V}_1$ is outside of $\mathbf{R}$. This yields Inside $(\mathbf{V}_1, \mathbf{R}) = \mathcal{F}$.

$\mathbf{V}_2$: $\mathbf{DetRes}_t^{\mathbf{V}_2}(\mathcal{S}_2)$ and thus $\mathbf{POP}_t^{\mathbf{V}_2} \subseteq \mathbf{R}^I$. Hence, Inside $(\mathbf{V}_2, \mathbf{R}) = \mathcal{T}$.

$\mathbf{V}_3$: Since the distance between $\mathcal{S}_3$ and the border of $\mathbf{R}$ is less than $\mathcal{D}_{max}$, the detection area could overlap the border. If a vehicle is detected only by $\mathcal{S}_3$, the SN cannot determine on which side of the border it is. Thus, Inside $(\mathbf{V}_3, \mathbf{R}) = \mathcal{M}$.

$\mathbf{V}_4$: Analogously to $\mathbf{V}_3$. ◆

The mapping of each detection scenario to a result for any predicate is specified in the following. We prove for each predicate P $(\mathbf{O}, \mathbf{R})$ that the set of objects $\mathbf{O}$ where P $(\mathbf{O}, \mathbf{R}) = \mathcal{M}$ is minimal, i.e., the result obtained this way is optimal. This mapping gives way to meaningful results for spatio-temporal developments in Section 7.

### 6.2.1 Deriving Results for Inside (O, R)

Considering the five detection scenarios, there are two scenarios where an object could be in a region $\mathbf{R}$ and one where this is certain:

$\mathsf{DS}^I$: $\mathbf{POP}_t^{\mathbf{O}}$ only intersects with $\mathbf{R}^I$, i.e., $\mathbf{POP}_t^{\mathbf{O}} \subseteq \mathbf{R}^I$. Hence, $\mathbf{O}$ is in $\mathbf{R}$ for sure.

$\mathsf{DS}^\bullet$: $\mathbf{POP}_t^{\mathbf{O}}$ overlaps with $\mathbf{R}^I$ but also overlaps with other partitions of $\mathbf{R}$. Thus, it is possible that $\mathbf{O}$ fulfills Inside $(\mathbf{O}, \mathbf{R})$ but is not guaranteed.

$\mathsf{DS}^\varnothing$: Objects may be in $\mathbf{R}$ without being detected, i.e., $\mathbf{O}$ might fulfill Inside $(\mathbf{O}, \mathbf{R})$ while being undetected.

Equation 22 summarizes the mapping of detection scenarios to predicate results for Inside $(\mathbf{O}, \mathbf{R})$:

$$\text{Inside}(\mathbf{O}, \mathbf{R}) = \begin{cases} \mathcal{T} \text{ iff } \mathsf{DS}^I \\ \mathcal{F} \text{ iff } \mathsf{DS}^E, \mathsf{DS}^B \\ \mathcal{M} \text{ iff } \mathsf{DS}^\bullet, \mathsf{DS}^\varnothing \end{cases} \tag{22}$$

**Lemma 4.** *Let $\Omega_{Inside}^{\mathbf{R}}$ be the set of objects in $\mathbf{R}$. The set of objects where Inside $(\mathbf{O}, \mathbf{R})$ yields $\mathcal{T}$ or $\mathcal{M}$ is the smallest superset of $\Omega_{Inside}^{\mathbf{R}}$ that the* SN *can derive.*

*Proof.* The lemma is true if the objects detected with $\mathsf{DS}^E$ and $\mathsf{DS}^B$ do not fulfill Inside $(\mathbf{O}, \mathbf{R})$ for sure. $\mathsf{DS}^E$ means that $\mathbf{POP}_t^{\mathbf{O}}$ is a subset of $\mathbf{R}^E$, i.e., $\mathbf{POP}_t^{\mathbf{O}}$ does not intersect with $\mathbf{R}^I$. The detection scenario $\mathsf{DS}^B$ occurs for objects that are on the border, i.e., $\mathbf{POP}_t^{\mathbf{O}}$ is a subset of $\mathbf{R}^B$. Hence, the object is not in $\mathbf{R}$ in both cases for sure. ■

**Lemma 5.** *The set of objects where Inside* $(\mathbf{O}, \mathbf{R}) = \mathcal{T}$ *is the largest subset of* $\Omega^{\mathbf{R}}_{Inside}$ *that the* SN *can derive.*

*Proof.* Only objects detected according to $\mathsf{DS}^I$ correspond to object that fulfill Inside $(\mathbf{O}, \mathbf{R})$ for sure. The remaining detection scenarios cannot guarantee that the detected object is in $\mathbf{R}$. $\mathsf{DS}^{\varnothing}$ and $\mathsf{DS}^{\bullet}$ may occur for objects outside of $\mathbf{R}$ as well. Objects detected according to $\mathsf{DS}^E$ or $\mathsf{DS}^B$ are not in $\mathbf{R}$ for sure. Thus, there does not exist a detection scenario of $\mathbf{O}$ that guarantees Inside $(\mathbf{O}, \mathbf{R})$ except $\mathsf{DS}^I$. ■

### 6.2.2 Deriving Results for Meet (O,R)

The predicate Meet $(\mathbf{O}, \mathbf{R})$ is true if $\mathbf{O}$ is on the border $\mathbf{R}^B$ of the region $\mathbf{R}$. From the set of detection scenarios, there is one that guarantees that $\mathbf{O}$ is on the border and two others where it is possible:

- $\mathsf{DS}^B$: In this case $\mathbf{POP}^{\mathbf{O}}_t \subseteq \mathbf{R}^B$, i.e., Meet $(\mathbf{O}, \mathbf{R}) = \mathcal{T}$.
- $\mathsf{DS}^{\bullet}$: In contrast to the previous case, $\mathbf{POP}^{\mathbf{O}}_t$ also contains points that are not part of the border. Thus, the object could be on the border, but the limited accuracy of the detection mechanism does not allow a definitive answer, i.e., Meet $(\mathbf{O}, \mathbf{R}) = \mathcal{M}$.
- $\mathsf{DS}^{\varnothing}$: The object could be on the border while not being detected by any sensor node, and therefore Meet $(\mathbf{O}, \mathbf{R}) = \mathcal{M}$ in this case.

Equation (23) summarizes this:

$$\text{Meet}\,(\mathbf{O}, \mathbf{R}) = \begin{cases} \mathcal{T} \text{ iff } \mathsf{DS}^B \\ \mathcal{F} \text{ iff } \mathsf{DS}^I, \mathsf{DS}^E \\ \mathcal{M} \text{ iff } \mathsf{DS}^{\bullet}, \mathsf{DS}^{\varnothing} \end{cases} \tag{23}$$

**Lemma 6.** *Let* $\Omega^{\mathbf{R}}_{Meet}$ *be the set of objects on the border* $\mathbf{R}^B$*. The set of objects where Meet* $(\mathbf{O}, \mathbf{R})$ *yields* $\mathcal{T}$ *or* $\mathcal{M}$ *is the smallest superset of* $\Omega^{\mathbf{R}}_{Meet}$ *that a* SN *can derive based on detection scenarios.*

*Proof.* Analogously to Lemma 4, we prove this by considering $\mathsf{DS}^I$ and $\mathsf{DS}^E$: $\mathsf{DS}^I$ ensures that $\mathbf{POP}^{\mathbf{O}}_t$ only contains points from $\mathbf{R}^I$, i.e., $\mathbf{O}$ is not on the border $\mathbf{R}^B$. Similarly, we derive from $\mathsf{DS}^E$ that $\mathbf{POP}^{\mathbf{O}}_t$ is a subset of $\mathbf{R}^E$ and thus does not intersect with $\mathbf{R}^B$. Thus, the set of objects where Meet $(\mathbf{O}, \mathbf{R})$ yields $\mathcal{T}$ or $\mathcal{M}$ is the smallest superset of $\Omega^{\mathbf{R}}_{Meet}$ the sensor network can compute. ■

**Lemma 7.** *The set of objects where Meet* $(\mathbf{O}, \mathbf{R}) = \mathcal{T}$ *is the largest subset of* $\Omega^{\mathbf{R}}_{Meet}$ *identifiable by the* SN.

*Proof.* Only $\mathsf{DS}^B$ yields Meet $(\mathbf{O}, \mathbf{R}) = \mathcal{T}$. Objects $\mathbf{O}$ detected according to $\mathsf{DS}^{\bullet}$ could be on $\mathbf{R}^B$, but it is not sure, because $\mathbf{POP}^{\mathbf{O}}_t$ also contains points from other partitions. Undetected objects could be on the border as well, but since they are not detected, it is not certain. For the other two detection scenarios, it is sure that the detected object is not on the border because $\mathbf{POP}_t^{\mathbf{O}} \cap \mathbf{R}^B = \varnothing$. Thus, a sensor network cannot compute a larger subset of $\Omega^{\mathbf{R}}_{Meet}$. ■

As stated above, most detection mechanisms used in SN cannot determine that some object $\mathbf{O}$ is on $\mathbf{R}^B$. Thus, once the distance of an object $\mathbf{O}$ to $\mathbf{R}^B$ falls below a certain limit, the detection mechanism cannot determine if the object is on the border or just close to it. Even if the sensor nodes can distinguish between stationary objects on $\mathbf{R}^B$ and those close to $\mathbf{R}^B$, the result of the detection would be $\mathsf{DS}^\bullet$ in most cases instead of $\mathsf{DS}^B$: The border $\mathbf{R}^B$ is a line. The time it takes for an object to move over this line is infinitely short. Capturing this moment reliably would require hardware with infinitely high temporal resolution. Thus, even with very sophisticated detection mechanisms, SN cannot detect objects on the border reliably.

Summing up, the set of objects detected with $\mathsf{DS}^B$ is typically very small or empty. But there are cases where a SN might be able to guarantee that an object is on the border and therefore we cannot ignore $\mathsf{DS}^B$. One might consider removing Meet $(\mathbf{O}, \mathbf{R})$ from the set of predicates for SN where it is impossible to detect an object with $\mathsf{DS}^B$, since the only case where Meet $(\mathbf{O}, \mathbf{R}) = \mathcal{T}$ will not occur. However, removing it is problematic as it would reduce the set of spatio-temporal queries expressible in SN significantly. For example, without Meet $(\mathbf{O}, \mathbf{R})$ one cannot express the development Touch $(\mathbf{O}, \mathbf{R})$. We show in Section 7 that there exist developments containing Meet $(\mathbf{O}, \mathbf{R})$ whose meaning can be guaranteed despite these problems. We conclude that the mapping in (23) for Meet $(\mathbf{O}, \mathbf{R})$ is as accurate as the detection mechanisms allow.

### 6.2.3 Deriving Results for Disjoint (O,R)

To conform to Disjoint $(\mathbf{O}, \mathbf{R})$, object $\mathbf{O}$ must be in $\mathbf{R}^E$. The mapping to detection scenarios is analogous to Inside $(\mathbf{O}, \mathbf{R})$:

$$\text{Disjoint}(\mathbf{O}, \mathbf{R}) = \begin{cases} \mathcal{T} \text{ iff } \mathsf{DS}^E \\ \mathcal{F} \text{ iff } \mathsf{DS}^I, \mathsf{DS}^B \\ \mathcal{M} \text{ iff } \mathsf{DS}^\bullet, \mathsf{DS}^\varnothing \end{cases} \tag{24}$$

There are lemmas analogous to Lemmas 4 and 5 for Disjoint $(\mathbf{O}, \mathbf{R})$. Hence, we conclude that the result in (24) is as accurate as possible as well.

### 6.2.4 Static and Dynamic Regions

The application scenarios in Section 2 have shown that there are static and dynamic regions. A *static region* $\mathbf{R}$ is a point set that does not change over time, while the point set representing a *dynamic region* does. The predicate results defined above apply to static and dynamic regions. Computing the detection scenario to obtain a predicate result implicitly assumes that the point set representing the region is known. Thus, prior to computing a detection scenario, it is necessary to determine which points $\mathsf{p} \in \mathbb{E}^d$ are in the region $\mathbf{R}$.

For a static region $\mathbf{R}$, computing a polygon encompassing $\mathbf{R}$ before query processing is straightforward. Each node can store the polygon and compute the intersection $\mathbf{POP}_t^{\mathbf{O}} \cap \mathbf{R}$, i.e., derive a detection scenario.

Checking if a point $\mathsf{p}$ is inside or outside of $\mathbf{R}$ becomes problematic if $\mathbf{R}$ is dynamic, i.e., changes over time. The problem is illustrated in Example 7.

**Example 7:** Suppose $\mathbf{R}$ is the point set that contains all points with a temperature below $0°C$. If $\mathcal{S}_i$ detects an object $\mathbf{O}$ at time $t$ and computes $\mathbf{POP}_t^{\mathbf{O}}$, it is not possible to intersect $\mathbf{POP}_t^{\mathbf{O}}$ with the partitions of $\mathbf{R}$: If $\mathcal{S}_i$ measures a temperature below $0°C$, it is not certain that $\mathbf{O}$ also is at a position where the temperature is less than $0°C$. Analogously, $\mathcal{S}_i$ cannot rule out that $\mathbf{O}$ is at a position where the temperature is below $0°C$. ◆

Solving the problem described in Example 7 requires restrictive assumptions regarding the SN: There must be at least one node that can check $\mathsf{C}_{\mathbf{R}}(\mathsf{p})$ for every $\mathsf{p} \in \mathbb{E}^d$. This implies that nodes must be equipped with sophisticated hardware that allows checking $\mathsf{C}_{\mathbf{R}}(\mathsf{p})$ for points $\mathsf{p}$ where no node has been deployed. For instance, infra-red cameras allow a node to determine the temperature in its vicinity. However, nodes equipped with these cameras must have considerably more computational power than those available today to process the images taken by the cameras. Additionally, the nodes must be deployed in such a way that there is at least one camera that can measure the temperature for any point in space at any time. Summing up, processing spatio-temporal queries targeting at the relationship of an object and a dynamic region has strict prerequisites. However, it is sufficient for most SN if the movement of an object is observed in relation to a zone. We now define the respective predicates and show how to derive results for them based on detection scenarios.

## *6.3 Predicate Results for Zones*

Section 5 has proposed a space partitioning induced by a given zone $\mathsf{Z}$, based on detection areas. Even if sensors cannot determine their detection areas, we can derive the partition of the zone where an object detected is located by using the following concept: If a node $\mathcal{S}_i \in \mathsf{Z}$ detects $\mathbf{O}$ at time $t$, the position estimate $\mathbf{DetRes}_t^{\mathbf{O}}(\mathcal{S}_i)$ intersects with $\mathsf{Z}^I$, i.e., $\mathbf{DetRes}_t^{\mathbf{O}}(\mathcal{S}_i) \cap \mathsf{Z}^I \neq \varnothing$. The actual position of $\mathbf{O}$ is either exclusively observed by nodes in $\mathsf{Z}$, or nodes inside and outside of $\mathsf{Z}$ observe it. Thus, the object is either in $\mathsf{Z}^I$ or in $\mathsf{Z}^B$. If there exists a node outside of $\mathsf{Z}$ that detects $\mathbf{O}$, $\mathbf{O}$ is located in $\mathsf{Z}^B$, otherwise it is in $\mathsf{Z}^I$. Summing up, one has to consider how the detection set $\mathsf{DetSet}_t^{\mathbf{O}}$ (cf. Definition 7) intersects with $\mathsf{Z}$ and $\overline{\mathsf{Z}}$ to determine how $\mathbf{POP}_t^{\mathbf{O}}$ intersects with the partitions of the zone, i.e., compute the corresponding detection scenario.

**Lemma 8.** *The intersection of* $\mathsf{Z}$ *and* $\mathsf{DetSet}_t^{\mathbf{O}}$ *determines the detection scenario for some object* $\mathbf{O}$ *at some time* $t$*:*

$$\mathsf{DetSet}_t^{\mathbf{O}} \cap \mathsf{Z} = \varnothing \wedge \mathsf{DetSet}_t^{\mathbf{O}} \cap \overline{\mathsf{Z}} \neq \varnothing \Rightarrow \mathbf{POP}_t^{\mathbf{O}} \subseteq \mathsf{Z}^E$$
$$\mathsf{DetSet}_t^{\mathbf{O}} \cap \mathsf{Z} \neq \varnothing \wedge \mathsf{DetSet}_t^{\mathbf{O}} \cap \overline{\mathsf{Z}} = \varnothing \Rightarrow \mathbf{POP}_t^{\mathbf{O}} \subseteq \mathsf{Z}^I$$
$$\mathsf{DetSet}_t^{\mathbf{O}} \cap \mathsf{Z} \neq \varnothing \wedge \mathsf{DetSet}_t^{\mathbf{O}} \cap \overline{\mathsf{Z}} \neq \varnothing \Rightarrow \mathbf{POP}_t^{\mathbf{O}} \subseteq \mathsf{Z}^B$$

*Proof.* We prove $\mathsf{DetSet}_t^{\mathbf{O}} \cap \mathsf{Z} = \varnothing \wedge \mathsf{DetSet}_t^{\mathbf{O}} \cap \overline{\mathsf{Z}} \neq \varnothing \Rightarrow \mathbf{POP}_t^{\mathbf{O}} \subseteq \mathsf{Z}^E$: The left-hand side of the implication means that only nodes in $\overline{\mathsf{Z}}$ detect $\mathbf{O}$, i.e, $\mathsf{DetSet}_t^{\mathbf{O}} \subseteq \overline{\mathsf{Z}}$. Hence, we prove $\mathsf{DetSet}_t^{\mathbf{O}} \subseteq \overline{\mathsf{Z}} \Rightarrow \mathbf{POP}_t^{\mathbf{O}} \subseteq \mathsf{Z}^E$ by contradiction[4], i.e., we have to prove that if $\mathbf{POP}_t^{\mathbf{O}}$ is not a subset of $\mathsf{Z}^E$ then $\mathsf{DetSet}_t^{\mathbf{O}}$ is not a subset of $\overline{\mathsf{Z}}$. Let $\mathcal{S}_i \in \mathsf{Z}$ detect $\mathbf{O}$ at $t$, i.e., $detect\,(\mathcal{S}_i, \mathbf{O}, t) = \mathcal{T}$. Thus, $\mathbf{O}$ is somewhere in $\mathbf{DA}_i$. Since $\mathbf{POP}_t^{\mathbf{O}}$ is the intersection of the detection areas of all nodes that detect $\mathbf{O}$ at $t$, $\mathbf{POP}_t^{\mathbf{O}}$ must contain at least one $\mathsf{p} \in \mathbf{DA}_i$. Hence, $\mathbf{POP}_t^{\mathbf{O}}$ is not a subset of $\mathsf{Z}^E$, because $\mathsf{Z}^E$ contains only points exclusively observed by nodes in $\overline{\mathsf{Z}}$. If $\mathbf{POP}_t^{\mathbf{O}}$ would not contain at least one $\mathsf{p} \in \mathbf{DA}_i$ then $detect\,(\mathcal{S}_i, \mathbf{O}, t) = \mathcal{F}$. Summing up, $\mathsf{DetSet}_t^{\mathbf{O}} \subseteq \overline{\mathsf{Z}}$ implies $\mathbf{POP}_t^{\mathbf{O}} \subseteq \mathsf{Z}^E$. The other two implications can be proven similarly. ■

Note that the right-hand side of each implication equals the formal expression associated with the detection scenarios $\mathsf{DS}^E$, $\mathsf{DS}^I$ and $\mathsf{DS}^B$ respectively.

**Lemma 9.** *In the context of a zone* $\mathsf{Z}$*,* $\mathbf{POP}_t^{\mathbf{O}}$ *can never intersect with more than one partition of* $\mathsf{Z}$*:*

$$\mathbf{POP}_t^{\mathbf{O}} \cap \mathsf{Z}^E \neq \varnothing \Rightarrow \mathbf{POP}_t^{\mathbf{O}} \subseteq \mathsf{Z}^E$$
$$\mathbf{POP}_t^{\mathbf{O}} \cap \mathsf{Z}^I \neq \varnothing \Rightarrow \mathbf{POP}_t^{\mathbf{O}} \subseteq \mathsf{Z}^I$$
$$\mathbf{POP}_t^{\mathbf{O}} \cap \mathsf{Z}^B \neq \varnothing \Rightarrow \mathbf{POP}_t^{\mathbf{O}} \subseteq \mathsf{Z}^B$$

*Proof.* We prove $\mathbf{POP}_t^{\mathbf{O}} \cap \mathsf{Z}^E \neq \varnothing \Rightarrow \mathbf{POP}_t^{\mathbf{O}} \subseteq \mathsf{Z}^E$: According to Definition 16, $\mathsf{Z}^E$ only contains points that are exclusively observed by nodes in $\overline{\mathsf{Z}}$. Hence, if $\mathbf{POP}_t^{\mathbf{O}}$ contains points from $\mathsf{Z}^E$, the object is at a position that is exclusively observed by nodes in $\overline{\mathsf{Z}}$. If there exists a node $\mathcal{S}_i \in \mathsf{Z}$ that detects $\mathbf{O}$, $\mathbf{POP}_t^{\mathbf{O}}$ does not intersect with $\mathsf{Z}^E$ anymore. The proofs for the remaining two implications are analogous. ■

Due to Lemma 9, $\mathsf{DS}^\bullet$ cannot occur with zones. Thus, we omit $\mathsf{DS}^\bullet$ for the definition of predicates which express the relationship between an object and a zone.

[4] To prove $A \Rightarrow B$ by contradiction, it is sufficient to prove $\overline{\mathrm{B}} \Rightarrow \overline{\mathrm{A}}$.

**Definition 24 (Disjoint $(\mathbf{O}, \mathsf{Z})$):** The object $\mathbf{O}$ conforms to Disjoint $(\mathbf{O}, \mathsf{Z})$ if $\mathbf{O}$ is exclusively detected by nodes in $\overline{\mathsf{Z}}$, i.e., if $\mathsf{DS}^E$ occurs (cf. Lemma 8):

$$\text{Disjoint}\,(\mathbf{O}, \mathsf{Z}) = \begin{cases} \mathcal{T} \text{ iff } \mathsf{DS}^E \\ \mathcal{F} \text{ otherwise} \end{cases} \tag{25}$$

□

**Definition 25 (Inside $(\mathbf{O}, \mathsf{Z})$):** The object $\mathbf{O}$ conforms to Inside $(\mathbf{O}, \mathsf{Z})$ if $\mathbf{O}$ is exclusively detected by nodes in $\mathsf{Z}$, i.e., if $\mathsf{DS}^I$ occurs (cf. Lemma 8):

$$\text{Inside}\,(\mathbf{O}, \mathsf{Z}) = \begin{cases} \mathcal{T} \text{ iff } \mathsf{DS}^I \\ \mathcal{F} \text{ otherwise} \end{cases} \tag{26}$$

□

**Definition 26 (Meet $(\mathbf{O}, \mathsf{Z})$):** The object $\mathbf{O}$ conforms to Meet $(\mathbf{O}, \mathsf{Z})$ if $\mathbf{O}$ is detected by nodes in $\mathsf{Z}$ and $\overline{\mathsf{Z}}$ simultaneously, i.e., if $\mathsf{DS}^B$ occurs (cf. Lemma 8):

$$\text{Meet}\,(\mathbf{O}, \mathsf{Z}) = \begin{cases} \mathcal{T} \text{ iff } \mathsf{DS}^B \\ \mathcal{F} \text{ otherwise} \end{cases} \tag{27}$$

□

Let $\Omega^{\mathsf{Z}}_{Disjoint}$ be the set of objects in $\mathsf{Z}^E$. Since there is no detection scenario where Disjoint $(\mathbf{O}, \mathsf{Z}) = \mathcal{M}$, we conclude that the set of objects where Disjoint $(\mathbf{O}, \mathsf{Z})$ yields $\mathcal{T}$ equals $\Omega^{\mathsf{Z}}_{Disjoint}$. Similarly, the sets of objects where Meet $(\mathbf{O}, \mathsf{Z})$ and Inside $(\mathbf{O}, \mathsf{Z})$ yield $\mathcal{T}$ equal $\Omega^{\mathsf{Z}}_{Meet}$ and $\Omega^{\mathsf{Z}}_{Inside}$ respectively.

The space partitioning for regions divides all points of space into three partitions. Every resulting partition is associated with a predicate. For zones, we have introduced a fourth partition $\mathsf{Z}^{\varnothing}$ which contains all points that are unobserved. To allow users to express that an object movement includes that the object is unobserved at some point in time, we define a fourth predicate:

**Definition 27 (Undetected $(\mathbf{O})$):** An object $\mathbf{O}$ conforms to Undetected $(\mathbf{O})$ if there is no node $\mathcal{S}_i \in \mathsf{SN}$ that detects $\mathbf{O}$:

$$\text{Undetected}\,(\mathbf{O}) = \begin{cases} \mathcal{T} \text{ iff } \mathsf{DS}^{\varnothing} \\ \mathcal{F} \text{ Otherwise} \end{cases} \tag{28}$$

□

In the following, we will write Undetected $(\mathbf{O})$ instead of Undetected $(\mathbf{O}, \mathsf{Z})$, because an object $\mathbf{O}$ with Undetected $(\mathbf{O}) = \mathcal{T}$ is undetected in relation to any other zone as well.

MOD-concepts like concatenation (cf. Definition 4) are applicable to the aforementioned predicates as well. Thus, one can construct developments that query the spatio-temporal relationship of objects and zones. For instance, one could define:

$$\begin{aligned} \text{Enter}\,(\mathbf{O}, \mathsf{Z}) = {} & \text{Disjoint}\,(\mathbf{O}, \mathsf{Z}) \triangleright \text{Meet}\,(\mathbf{O}, \mathsf{Z}) \\ & \triangleright \text{Inside}\,(\mathbf{O}, \mathsf{Z}) \end{aligned} \tag{29}$$

Undetected $(\mathbf{O})$ is particularly useful in the context of spatio-temporal developments. For example, a user could be interested in objects that fulfill

Inside $(\mathbf{O}, \mathsf{Z})$ first and then move into an unobserved area:

$$\text{Disappear}(\mathbf{O}, \mathsf{Z}) = \text{Inside}(\mathbf{O}, \mathsf{Z}) \triangleright \text{Undetected}(\mathbf{O}) \tag{30}$$

Further examples for the use of this predicate are provided in Section 7 where spatio-temporal developments in sensor networks are discussed.

### 6.3.1 Static and Dynamic Zones

As with regions, there are dynamic and static zones. Users define a static zone Z by providing a set of conditions such that the set of nodes fulfilling it does not change over time. A dynamic zone changes over time, typically because it depends on a measurable value, e.g., a temperature threshold. As with regions, the predicates defined above are applicable in both cases.

Recall that dynamic regions have resulted in extremely strict requirements regarding the capabilities and deployment of nodes. Dynamic zones do not have such requirements, because every node only has to determine if it is inside the zone or not. For example, the dynamic zone in Table 1 requires each node to determine at certain points of time if it measures a temperature below $0°C$. Measuring the temperature is a standard feature of sensor nodes available, e.g., Sun SPOT sensor nodes [50]. We conclude that commercially available sensor nodes can deal with dynamic zones, but not necessarily with dynamic regions.

## *6.4 Summary*

Table 2 summarizes the mapping of detection scenarios to results of predicates expressing the relation between objects and regions in SN. Each row corresponds to a predicate and every column to a detection scenario that describes how $\mathbf{POP}_t^{\mathbf{O}}$ overlaps with the partitions of the region $\mathbf{R}$.

| $\mathsf{P}(\mathbf{O}, \mathbf{R})$ | $\mathsf{DS}^{\varnothing}$ | $\mathsf{DS}^{E}$ | $\mathsf{DS}^{I}$ | $\mathsf{DS}^{B}$ | $\mathsf{DS}^{\bullet}$ |
|---|---|---|---|---|---|
| Inside $(\mathbf{O}, \mathbf{R})$ | $\mathcal{M}$ | $\mathcal{F}$ | $\mathcal{T}$ | $\mathcal{F}$ | $\mathcal{M}$ |
| Meet $(\mathbf{O}, \mathbf{R})$ | $\mathcal{M}$ | $\mathcal{F}$ | $\mathcal{F}$ | $\mathcal{T}$ | $\mathcal{M}$ |
| Disjoint $(\mathbf{O}, \mathbf{R})$ | $\mathcal{M}$ | $\mathcal{T}$ | $\mathcal{F}$ | $\mathcal{F}$ | $\mathcal{M}$ |

**Table 2** Mapping detection scenarios to predicate results for an object **O** and a region **R**

Predicates that describe the relation between an object and a zone are summarized similarly in Table 3. Since $\mathsf{DS}^{\bullet}$ cannot occur in the context of zones, the corresponding column contains '-' entries. Based on these results, we now focus on spatio-temporal developments, i.e., sequences of predicates that describe an object movement in relation to a query context.

| P ($\mathbf{O}$, Z) | $\mathsf{DS}^{\varnothing}$ | $\mathsf{DS}^{E}$ | $\mathsf{DS}^{I}$ | $\mathsf{DS}^{B}$ | $\mathsf{DS}^{\bullet}$ |
|---|---|---|---|---|---|
| Inside ($\mathbf{O}$, Z) | $\mathcal{F}$ | $\mathcal{F}$ | $\mathcal{T}$ | $\mathcal{F}$ | - |
| Meet ($\mathbf{O}$, Z) | $\mathcal{F}$ | $\mathcal{F}$ | $\mathcal{F}$ | $\mathcal{T}$ | - |
| Disjoint ($\mathbf{O}$, Z) | $\mathcal{F}$ | $\mathcal{T}$ | $\mathcal{F}$ | $\mathcal{F}$ | - |
| Undetected ($\mathbf{O}$,Z) | $\mathcal{T}$ | $\mathcal{F}$ | $\mathcal{F}$ | $\mathcal{F}$ | - |

**Table 3** Mapping detection scenarios to predicate results for an object $\mathbf{O}$ and a zone Z

# 7 Spatio-Temporal Developments

As illustrated in Section 3.2.2, users express queries through spatio-temporal developments, i.e., by concatenating predicates. One core contribution of this paper is the translation of sequences of object detections to results for spatio-temporal developments.

There are some preliminary steps for such a translation: First, we show that the concatenation operator $\triangleright$ (cf. Definition 4) is insufficient to express certain information needs in SN. We address this by introducing a new concatenation operator. Second, we develop a canonical collection of spatio-temporal developments for SN similar to the existing collection for moving object databases [19]. We need this collection to obtain a finite set of developments which we must translate to sequences of object detections. The last step is the actual translation of each element of the canonical collection and a proof that this translation is correct.

## *7.1 Irregularity of Zones and Concatenation*

The difference between the partitioning of space for regions and the one for zones is that regularity [51] cannot be assumed for zones: Among other things, regularity means that the interior $\mathbf{R}^I$ is completely encompassed by the border $\mathbf{R}^B$ of a region $\mathbf{R}$. As shown in Figure 8, this is different with zones: The interior $Z^I$ adjoins to the border $Z^B$ and $Z^{\varnothing}$.

The semantics of developments like Enter ($\mathbf{O}$, $\mathbf{R}$) are affected by this: Suppose that a user is interested in all objects $\mathbf{O}$ that move into the zone Z. For regions, the space partitions are regular, i.e., an object $\mathbf{O}$ must cross the border $\mathbf{R}^B$. In the context of a zone, a user could express an interest similar to

Enter $(\mathbf{O}, \mathbf{R})$ with Enter $(\mathbf{O}, \mathrm{Z})$, as defined in (29). This is problematic, because Enter $(\mathbf{O}, \mathrm{Z})$ restricts the result to objects that are observed explicitly while crossing the border. However, an object $\mathbf{O}$ might fulfill Disjoint $(\mathbf{O}, \mathrm{Z})$ at some time, then move through an unobserved area and fulfill Inside $(\mathbf{O}, \mathrm{Z})$ afterward. From a semantical perspective, $\mathbf{O}$ has entered the zone, but does not fulfill Enter $(\mathbf{O}, \mathrm{Z})$.

One might solve this by querying for all objects that either fulfill Enter $(\mathbf{O}, \mathrm{Z})$ or HiddenEnter $(\mathbf{O}, \mathrm{Z})$, which is defined in (31):

$$\begin{aligned}\text{HiddenEnter}(\mathbf{O}, \mathrm{Z}) = \text{Disjoint}(\mathbf{O}, \mathrm{Z}) &\rhd \text{Undetected}(\mathbf{O}) \\ &\rhd \text{Inside}(\mathbf{O}, \mathrm{Z})\end{aligned} \tag{31}$$

HiddenEnter $(\mathbf{O}, \mathrm{Z})$ is insufficient as well: $\mathbf{O}$ could fulfill Disjoint $(\mathbf{O}, \mathrm{Z})$ first, then Undetected $(\mathbf{O})$ followed by Meet $(\mathbf{O}, \mathrm{Z})$ and finally Inside $(\mathbf{O}, \mathrm{Z})$. In this case, $\mathbf{O}$ neither fulfills HiddenEnter $(\mathbf{O}, \mathrm{Z})$ nor Enter $(\mathbf{O}, \mathrm{Z})$. A user with the aforementioned query who does not care if the object is detected or not while crossing the border would have to provide an infinite number of predicate sequences. This is because an object can move an arbitrary number of times between Undetected $(\mathbf{O})$ and Meet $(\mathbf{O}, \mathrm{Z})$ before fulfilling Inside $(\mathbf{O}, \mathrm{Z})$. The development in (32) is not an option either:

$$\text{Disjoint}(\mathbf{O}, \mathrm{Z}) \rhd \text{Inside}(\mathbf{O}, \mathrm{Z}) \tag{32}$$

The sequence in (32) never occurs, because $\rhd$ requires Inside $(\mathbf{O}, \mathrm{Z})$ to follow Disjoint $(\mathbf{O}, \mathrm{Z})$ immediately.

**Lemma 10.** *For any object* $\mathbf{O}$ *and a region* $\mathbf{R}$*, there does not exist a movement that fulfills Inside* $(\mathbf{O}, \mathbf{R}) \rhd$ *Disjoint* $(\mathbf{O}, \mathbf{R})$*. Objects cannot fulfill Disjoint* $(\mathbf{O}, \mathbf{R}) \rhd$ *Inside* $(\mathbf{O}, \mathbf{R})$ *as well.*

*Proof.* According to Definition 4, the movement of an object $\mathbf{O}$ in relation to a region $\mathbf{R}$ satisfies Inside $(\mathbf{O}, \mathbf{R}) \rhd$ Disjoint $(\mathbf{O}, \mathbf{R})$ if Inside $(\mathbf{O}, \mathbf{R}) = \mathcal{T}$ for some interval $[t_0, t_1[$ and Disjoint $(\mathbf{O}, \mathbf{R}) = \mathcal{T}$ at $t_1$. Due to the partitioning of space defined for regions (cf. Section 3.2), to satisfy Inside $(\mathbf{O}, \mathbf{R})$ at $t_i$ and Disjoint $(\mathbf{O}, \mathbf{R})$ later at $t_j$, the object must cross the border at $t_i < t < t_j$. Thus, if Inside $(\mathbf{O}, \mathbf{R}) = \mathcal{T}$ for $[t_0, t_1[$, Meet $(\mathbf{O}, \mathbf{R}) = \mathcal{T}$ at $t_1$. Hence, Disjoint $(\mathbf{O}, \mathbf{R})$ is not possible at $t_1$. For Disjoint $(\mathbf{O}, \mathbf{R}) \rhd$ Inside $(\mathbf{O}, \mathbf{R})$, the proof is analogous. ■

**Definition 28 (Relaxed Concatenation):** The *relaxed concatenation of two predicates*, P $\tilde{\rhd}$ Q, is true if P is true for some time interval $[t_0; t_1[$, and Q is true at $t_2 \geq t_1$. □

Equation (33) defines a development that expresses the query discussed above:

$$\text{SNEnter}(\mathbf{O}, \mathrm{Z}) = \text{Disjoint}(\mathbf{O}, \mathrm{Z}) \,\tilde{\rhd}\, \text{Inside}(\mathbf{O}, \mathrm{Z}) \tag{33}$$

In combination with the predicate Undetected ($\mathbf{O}$), this new operator increases the semantical depth. Users now can explicitly define if the object must be observed or not while moving, as illustrated next.
**Example 8:** Figure 2 shows a SN deployed close to a river with several bridges. Suppose that nodes are deployed in a controlled way so that caribous moving over a bridge are detected, but caribous swimming are not, i.e., the river itself is unobserved. A user only interested in caribous **C** entering Z by crossing bridges can use Enter (**C**, Z). If only caribous that enter Z by swimming are of interest, the user can express this with HiddenEnter (**C**, Z). A user interested in all caribous entering Z can query SNEnter (**C**, Z). ◆

**Lemma 11.** $P_1 \triangleright P_2 \Rightarrow P_1 \tilde{\triangleright} P_2$

*Proof.* According to Definition 28, the right-hand side is true if $P_1$ is true for some interval $[t_0, t_1[$ and $P_2$ is true at $t_2 \geq t_1$. The left-hand side of the implication states that $P_1$ is true for some interval $[t_0, t_1[$ and $P_2$ is true at $t_2 = t_1$. Hence, if the left-hand side is true, the right-hand side is true as well. ■

**Lemma 12.** $P_1 \tilde{\triangleright} (P_2 \tilde{\triangleright} P_3) = (P_1 \tilde{\triangleright} P_2) \tilde{\triangleright} P_3$.

*Proof.* The left-hand side means $\exists [t_0, t_1[ \; : \; P_1$ and $\exists t_2 \geq t_1 \; : \; (P_2 \tilde{\triangleright} P_3)$. Furthermore, $\exists [t_2, t_3[ \; : P_2$ and $\exists t_4 \geq t_3 : P_3$. The right-hand side expresses that $\exists \left[t_0', t_3'\right[ \; : (P_1 \tilde{\triangleright} P_2)$ and $\exists t_4' \geq t_3' : P_3$. Additionally, $\exists \left[t_0', t_1'\right[, t_1' \leq t_3' :$ $P_1$ and $\exists t_2' \geq t_1' \wedge t_2' \leq t_3' : P_2$. If the left-hand side is true for $t_0' = t_0, t_1' = t_1, t_2' = t_2, t_3' = t_3$ the right-hand side is fulfilled also (and vice versa). ■

**Lemma 13.** $P_1 \triangleright (P_2 \tilde{\triangleright} P_3) = (P_1 \triangleright P_2) \tilde{\triangleright} P_3$

*Proof.* By applying Lemma 11, we derive that $P_1 \triangleright (P_2 \tilde{\triangleright} P_3)$ implies $P_1 \tilde{\triangleright} (P_2 \tilde{\triangleright} P_3)$. Analogously applying Lemma 11 to the right-hand side of the implication results in $(P_1 \tilde{\triangleright} P_2) \tilde{\triangleright} P_3$. Thus, we get $P_1 \tilde{\triangleright} (P_2 \tilde{\triangleright} P_3) = (P_1 \tilde{\triangleright} P_2) \tilde{\triangleright} P_3$ which is true according to Lemma 12. ■

Users can formulate queries using both concatenation operators. Thus, we define spatio-temporal developments in the context of SN as follows:
**Definition 29 (Spatio-Temporal Development):** A *spatio-temporal development* $\mathbb{P}$ is a sequence of predicates $\mathbb{P} = P_1 \; \theta \; P_2 \; \theta \; \ldots \; \theta \; P_q$ with $\theta \in \{\triangleright, \tilde{\triangleright}\}$. The movement of an object $\mathbf{O}$ conforms to $\mathbb{P}$ if each pair $P_{i-1} \; \theta \; P_i$ with $2 \leq i \leq q$ is true in the order defined by $\mathbb{P}$. □

We denote developments that describe the relation of an object $\mathbf{O}$ and a region $\mathbf{R}$ with $\mathbb{P}(\mathbf{O}, \mathbf{R})$. In this case, all predicates refer to $\mathbf{O}$ and $\mathbf{R}$ as well, i.e., $P_i = P_i(\mathbf{O}, \mathbf{R})$ with $1 \leq i \leq q$. Similarly, $\mathbb{P}(\mathbf{O}, Z)$ describes the spatio-temporal relationship of $\mathbf{O}$ and a zone Z.

We use this definition to derive a canonical collection of developments for SN. This collection limits the set of developments which must be translated into sequences of object detections.

## *7.2 A Canonical Collection of Spatio-Temporal Developments*

To obtain a canonical collection of spatio-temporal developments, [19] constructs a *development graph* which represents possible spatio-temporal developments. A development is possible if an object can move such that the corresponding sequence of predicates $\mathrm{P}_1 \;\theta\, \mathrm{P}_2 \;\theta \ldots \;\theta\; \mathrm{P}_q$ is satisfied.

**Definition 30 (Development Graph):** A *development graph* is a graph DG = (V, E) that expresses possible predicate sequences:

V: Each possible predicate is represented by a vertex.

E: There is an edge $(\mathrm{P}_i, \mathrm{P}_j)$ if an object can move such that $\mathrm{P}_i \;\theta\; \mathrm{P}_j$ is satisfied. □

As shown above, the set of predicates applicable to regions and objects differs from the one for zones and objects: While there are equivalents to Inside $(\mathbf{O}, \mathbf{R})$, Meet $(\mathbf{O}, \mathbf{R})$ and Disjoint $(\mathbf{O}, \mathbf{R})$, the set of predicates for zones also contains Undetected $(\mathbf{O})$. Thus, the development graph for zones is different from the one for regions.

### 7.2.1 The Object/Region Development Graph

The set of vertices $\mathrm{V}^{\mathbf{R}}$ of the object/region development graph $\mathrm{DG}^{\mathbf{R}} = (\mathrm{V}^{\mathbf{R}}, \mathrm{E}^{\mathbf{R}})$ has three elements:

$$\mathrm{V}^{\mathbf{R}} = \{\mathrm{Inside}\,(\mathbf{O}, \mathbf{R}),\ \mathrm{Meet}\,(\mathbf{O}, \mathbf{R}),\ \mathrm{Disjoint}\,(\mathbf{O}, \mathbf{R})\}$$

Lemma 10 implies that there does not exist an edge from Disjoint $(\mathbf{O}, \mathbf{R})$ to Inside $(\mathbf{O}, \mathbf{R})$ and vice versa. Figure 10 shows the object/region development graph. For all graphs that follow, we use different lines to distinguish between the different concatenation operators: Solid lines represent concatenations that exist for both operators $\rhd$ and $\tilde{\rhd}$. The dotted lines stand for concatenations only possible with $\tilde{\rhd}$. Similarly, dashed lines represent concatenations with $\rhd$.

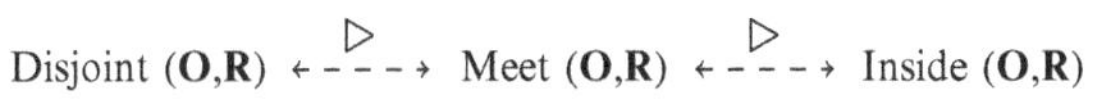


**Fig. 10** Development Graph for an object **O** and a region **R**

Comparing this graph to the development graph in Figure 11 for objects and regions in MOD shows that they only differ in one vertex: MOD distinguish between meet $(\mathbf{o}, \mathbf{r})$ and Meet $(\mathbf{O}, \mathbf{R})$ [19]. meet $(\mathbf{o}, \mathbf{r}) = \mathcal{T}$ if $\mathbf{o}$

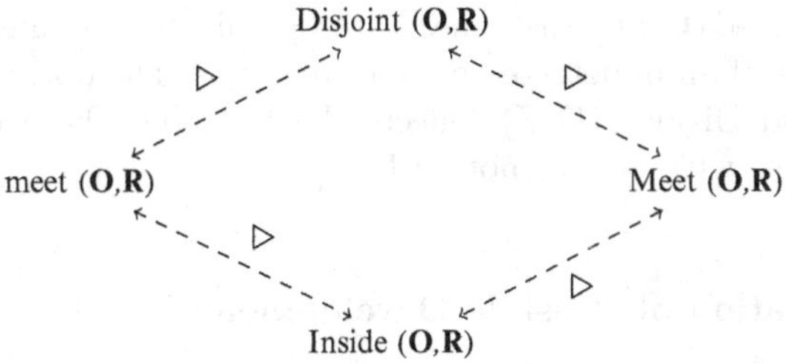


**Fig. 11** Development Graph for an object **O** and a region **R** in MOD according to [19]

is on the border of **R** for exactly one instant of time. Contrary to that, $\text{Meet}(\mathbf{O},\mathbf{R}) = \mathcal{T}$ if **O** is on the border of **R** for a time interval. We omit developments with $\text{meet}(\mathbf{o},\mathbf{r})$ for SN, since this would assume detection mechanisms with infinite temporal resolution.

### 7.2.2 The Object/Zone Development Graph

As shown in Section 6.3, there are four predicates that express the relationship between an object and a zone. Thus, for the object/zone development graph $\text{DG}^{\text{Z}} = \left(\text{V}^{\text{Z}}, \text{E}^{\text{Z}}\right)$, the set of vertices $\text{V}^{\text{Z}}$ contains the four predicates $\text{Inside}(\mathbf{O},\text{Z})$, $\text{Meet}(\mathbf{O},\text{Z})$, $\text{Undetected}(\mathbf{O})$ and $\text{Disjoint}(\mathbf{O},\text{Z})$. Figure 12 shows the development graph for an object and a zone.

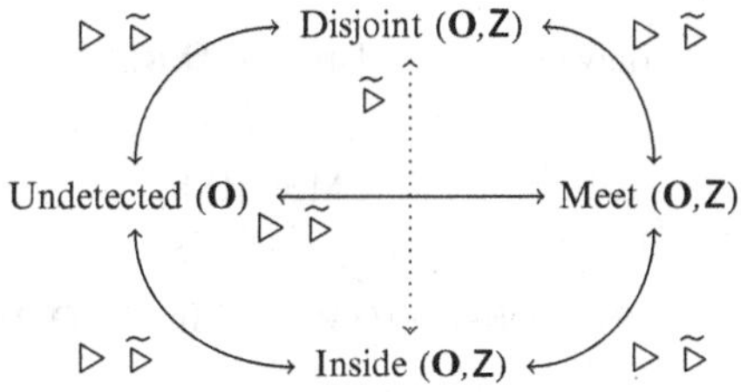


**Fig. 12** Development graph for an object **O** and a zone Z

Contrary to regions, zones are not regular (cf. Section 7.1). As we have shown, this irregularity necessitates the usage of two different concatenation operators. The edges in Figure 12 are explained as follows: The reasoning for the edges between $\text{Inside}(\mathbf{O},\text{Z})$, $\text{Meet}(\mathbf{O},\text{Z})$ and $\text{Disjoint}(\mathbf{O},\text{Z})$ is analogous to Section 7.2.1. Contrary to the development graph for regions, edges in Figure 10 are *solid*, i.e., they represent $\rhd$ and $\widetilde{\rhd}$. This is correct, because Lemma 11 has shown that $\text{P}_1 \rhd \text{P}_2 \Rightarrow \text{P}_1 \widetilde{\rhd} \text{P}_2$. Additional solid edges connect Undetected $(\mathbf{O})$ to the other three predicates, because objects can move into or out of an undetected area at any time. The dotted line between Inside $(\mathsf{O}, \mathsf{Z})$ and Disjoint $(\mathsf{O}, \mathsf{Z})$ reflects the fact that these predicates are only concatenable with $\widetilde{\triangleright}$, but not with $\triangleright$.

### 7.2.3 Enumeration of Possible Developments

Every path through a development graph represents a possible development. The number of these paths is infinite, due to cycles. Hence, one has to restrict the set of paths to obtain a finite set of developments. Similarly to [19], we obtain such a finite set by constructing development trees as follows:

1. Pick each element in V as the root of a development tree.
2. Generate a child node of this root for every vertex connected to this element in the development graph.
3. For each child node, construct a set of child nodes – the adjacent vertices in the development graph.
4. A node is a leaf node, i.e., node generation stops if
   a. every predicate exists on the path from the root to the current node, or
   b. the predicate corresponding to the current node already appears on the path from the root to the current node, i.e., there is a cycle.

To obtain the canonical collection, we generate all these trees based on the respective development graph. For regions, each node in such a tree represents one spatio-temporal development. As we show, a node in the trees for zones may represent more than one development.

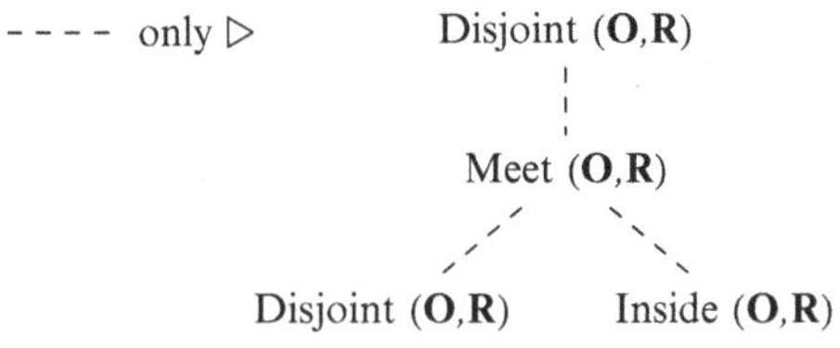


**Fig. 13** Development tree with root Disjoint $(\mathbf{O}, \mathbf{R})$

Figures 13-15 show the development trees with roots Disjoint $(\mathbf{O}, \mathbf{R})$, Meet $(\mathbf{O}, \mathbf{R})$ and Inside $(\mathbf{O}, \mathbf{R})$ respectively. The sum of nodes in these three trees is 13, i.e., there are 13 unique spatio-temporal developments that describe the relationship of an object and a region in a SN over time. These 13 developments include three developments consisting of a single predicate. Semantics of single predicates have been the focus of Section 6.2 already. The left column of Table 4 shows the ten developments consisting of more than one predicate. Section 7.4 shows how to derive results for these developments.

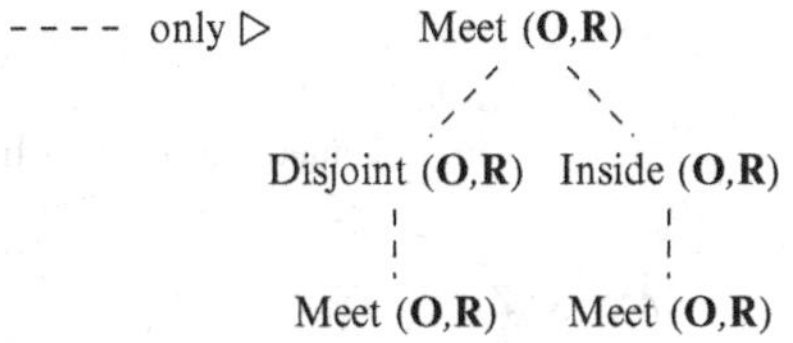


**Fig. 14** Development tree with root Meet $(\mathbf{O}, \mathbf{R})$

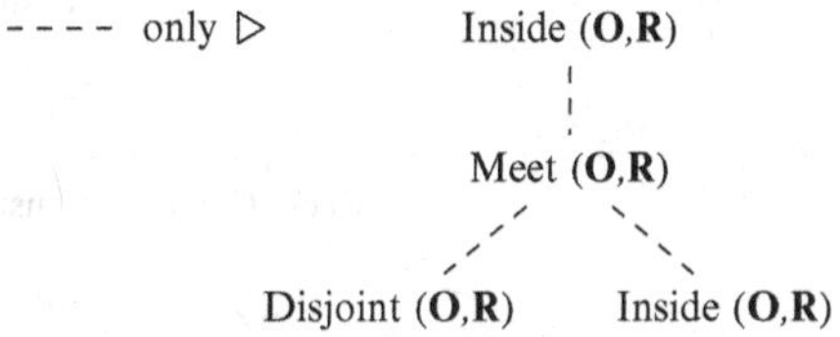


**Fig. 15** Development tree with root Inside $(\mathbf{O}, \mathbf{R})$

Figures 16-19 show the development trees for developments related to zones: Each tree has 31 nodes, i.e., the total number of nodes in all trees is $4 \cdot 31 = 124$. Contrary to the object/region development tree, each node represents more than one unique development because solid lines may be either $\rhd$ or $\widetilde{\rhd}$. The value above each node in Figure 16 indicates the number of developments represented by the node.

**Lemma 14.** *Every development tree related to zones represents* 146 *unique spatio-temporal developments.*

*Proof.* The sum of the numbers above the vertices of each development tree is 147. The value above each root vertex is 1, but contrary to all other vertices, this node does not represent a development, since it only represents a single predicate. Hence, to obtain the number of developments represented by the tree, one has to subtract 1 from the sum of the numbers above the vertices. The lemma holds if the number above every non-root vertex equals the number of developments represented by it. In the following, we suppose that the number above $v_i$ is $k_i$.

If $v_i$ is connected to $v_j$ via a solid edge, then $k_j = 2 \cdot k_i$. The vertex $v_i$ represents a set of $k_i$ predicate sequences that end with the predicate $\mathrm{P}_i$ associated with the vertex $v_i$. Since the edge between $v_i$ and $v_j$ is solid, it is possible to concatenate $\mathrm{P}_i$ with $\mathrm{P}_j$ using $\rhd$ and $\widetilde{\rhd}$. Thus, one can "append" $\mathrm{P}_j$ to each of these $k_i$ predicate sequences using either $\rhd$ or $\widetilde{\rhd}$. Hence, we conclude that the vertex $v_j$ represents $2 \cdot k_i$ developments that end with $\mathrm{P}_j$.

If $v_i$ is connected to $v_j$ via a dotted edge, then $k_j = k_i$. Again, $v_i$ represents a set of $k_i$ predicate sequences that end with the predicate $\mathrm{P}_i$ associated with the vertex $v_i$. Contrary to the case above, the dotted edge indicates that one

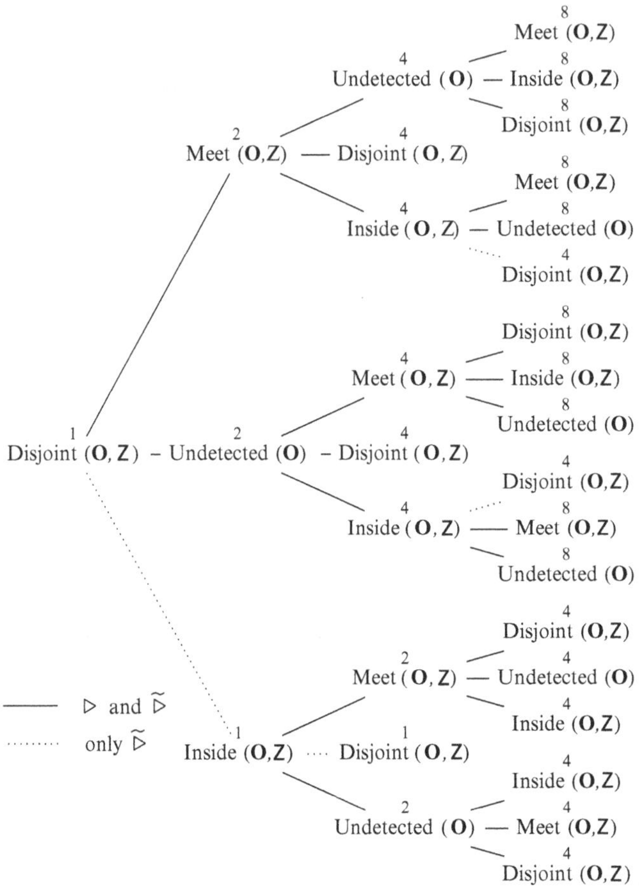


**Fig. 16** Development tree with root Disjoint (O, Z)

can only concatenate $\mathrm{P}_j$ to each of these $k_i$ predicate sequences using $\widetilde{\triangleright}$. Hence, $v_j$ represents $k_i$ developments that end with $\mathrm{P}_j$.

Summing up, the number above each node $v_i$ equals the number of developments represented by the path from the root node to $v_i$. Hence, we obtain the 146 spatio-temporal developments represented by every path in these development trees. ∎

We illustrate Lemma 14 using Figure 16: The root Disjoint $(\mathbf{O}, \mathrm{Z})$ in Figure 16 has edges to three predicates Undetected $(\mathbf{O})$, Inside $(\mathbf{O}, \mathrm{Z})$ and Meet $(\mathbf{O}, \mathrm{Z})$. The edge between Disjoint $(\mathbf{O}, \mathrm{Z})$ and Meet $(\mathbf{O}, \mathrm{Z})$ is solid, i.e., both predicates may be concatenated using $\triangleright$ and $\widetilde{\triangleright}$. Thus, there are two developments represented by this path:

1. Disjoint $(\mathbf{O}, \mathrm{Z}) \triangleright$ Meet $(\mathbf{O}, \mathrm{Z})$
2. Disjoint $(\mathbf{O}, \mathrm{Z}) \,\widetilde{\triangleright}\,$ Meet $(\mathbf{O}, \mathrm{Z})$.

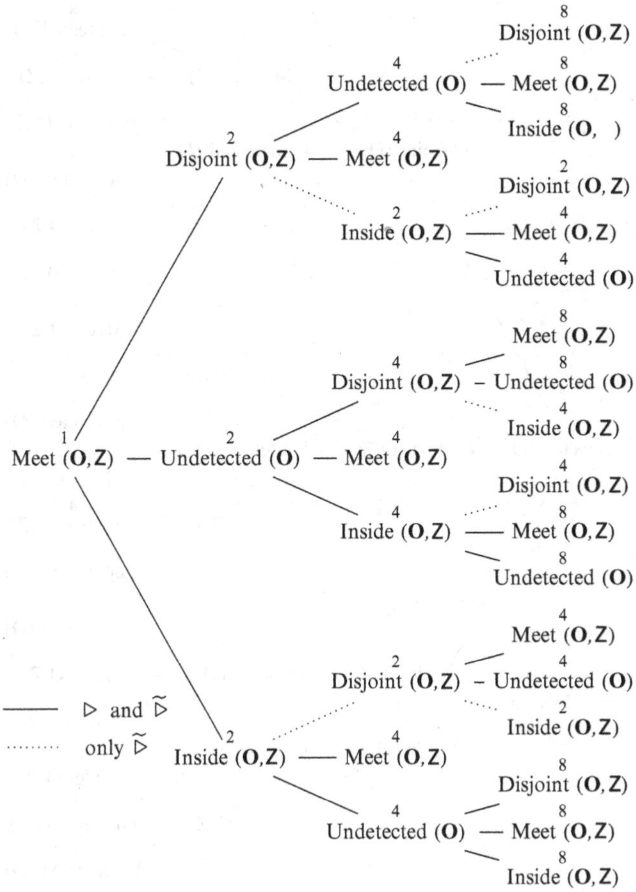


**Fig. 17** Development tree with root Meet $(\mathbf{O}, \mathrm{Z})$

The path from Disjoint $(\mathbf{O}, \mathrm{Z})$ to Undetected $(\mathbf{O})$ via Meet $(\mathbf{O}, \mathrm{Z})$ represents four developments. This is because one can "append" Undetected $(\mathbf{O})$ to each of the two developments above using either $\triangleright$ or $\widetilde{\triangleright}$.

The edge between the root node Disjoint $(\mathbf{O}, \mathrm{Z})$ and Inside $(\mathbf{O}, \mathrm{Z})$ is dotted. Thus, this path represents a single spatio-temporal development:

$$\text{Disjoint}\,(\mathbf{O}, \mathrm{Z}) \,\widetilde{\triangleright}\, \text{Inside}\,(\mathbf{O}, \mathrm{Z})$$

While the structure of the trees in Figures 16-19 varies slightly, each tree represents 146 unique spatio-temporal developments. Hence, users can express $4 \cdot 146 = 584$ unique spatio-temporal developments that describe the relationship between an object and a zone over time.

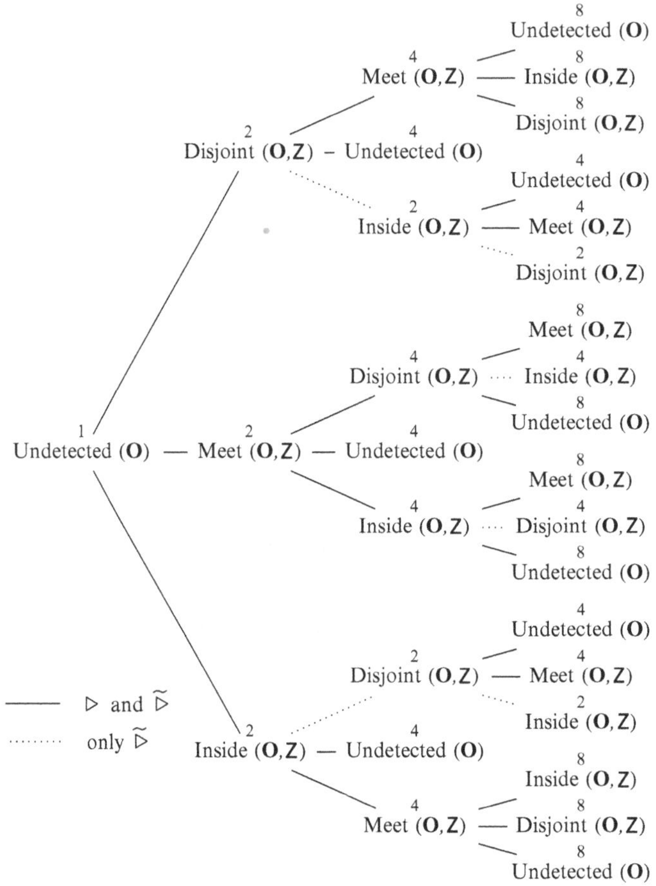


**Fig. 18** Development tree with root Undetected (**O**)

## *7.3 Formal Description of Object Detection Sequences*

The trajectory of on object matches a development if the object fulfills the predicates in the order specified by the development. We use the following operator to describe object trajectories formally:

**Definition 31 (Detection Concatenation):** The *concatenation of two detection scenarios*, $\mathsf{DS}_1 \succ \mathsf{DS}_2$, expresses that an object was detected according to $\mathsf{DS}_1$ in the time period $[t_1, t_2[$ and detected according to $\mathsf{DS}_2$ at $t_2$.[5] □

[5] We have chosen right-open intervals here to be in line with the definition of predicate sequences and the concatenation operator ▷ (cf. Definition 4). This does not cause any problems since the temporal resolution of any detection mechanism is limited in any case.

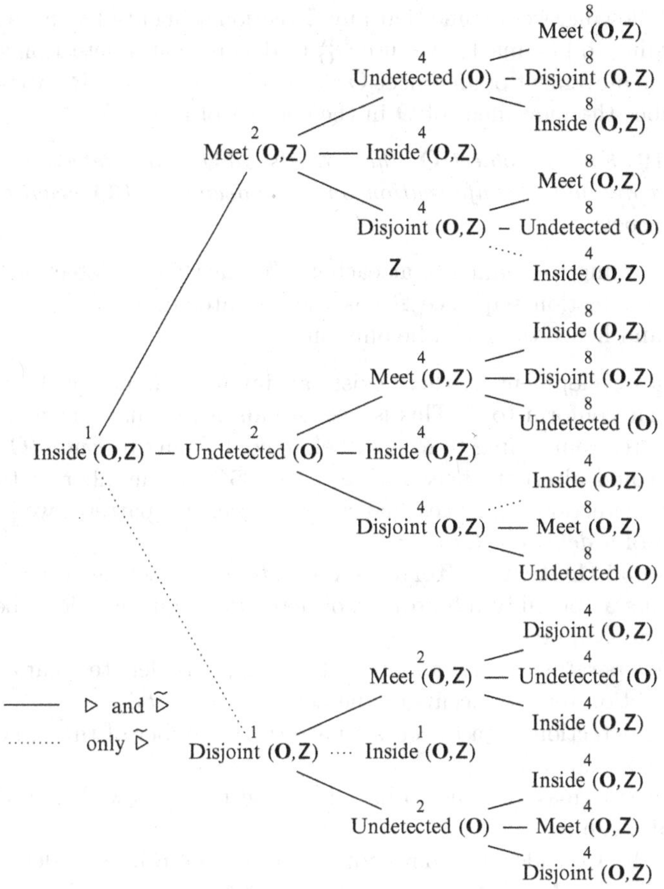


**Fig. 19** Development tree with root Inside (O, Z)

**Lemma 15.** $\mathsf{DS}_1 \succ \mathsf{DS}_1 = \mathsf{DS}_1$

*Proof.* The left-hand side means that there is an interval $[t_1, t_2[$ where an object is detected according to $\mathsf{DS}_1$ and another interval $[t_2, t_3[$ where the object is detected according to $\mathsf{DS}_1$ as well. This means that the object is detected according to $\mathsf{DS}_1$ during $[t_1, t_3[$ which equals the right-hand side. ■

**Definition 32 (Detection Sequence):** A *detection sequence* $\mathbb{D} = \mathsf{DS}_1 \succ \ldots \succ \mathsf{DS}_k$ is a concatenation of detection scenarios. It formalizes the information on the movement of an object with regard to a query context. $\mathbb{D}$ means that $\mathsf{DS}_1$ occurred for some time interval $[t_1, t_2[$, $\mathsf{DS}_2$ occurred for some interval $[t_2, t_3[$ etc. □

In the following, we assume that any detection sequence has been normalized according to Lemma 15. We use $\mathbb{D}_{\mathbf{O}}^{\mathbf{R}}$ to denote that a detection sequence refers to the movement of an object $\mathbf{O}$ in relation to a region $\mathbf{R}$. Analogously, $\mathbb{D}_{\mathbf{O}}^{\mathsf{Z}}$ describes the movement of $\mathbf{O}$ in the context of a zone $\mathsf{Z}$.

**Lemma 16.** *For any object* $\mathbf{O}$, *there exists exactly one detection sequence* $\mathbb{D}_{\mathbf{O}}$ *that represents the information on the movement of* $\mathbf{O}$ *acquired by the sensor network.*

*Proof.* According to Lemma 3, at each $t \in \mathbb{T}$ exactly one detection scenario holds. The detection sequence $\mathbb{D}_{\mathbf{O}}$ is the concatenation of these detection scenarios and hence there can be only one. ■

Given a development $\mathbb{P}$, there exists an infinite number of detection sequences that conform to $\mathbb{P}$. This is because an object may move arbitrarily before or after conforming to $\mathbb{P}$, e.g., before conforming to Enter $(\mathbf{O}, \mathbf{R})$, the object $\mathbf{O}$ could alternate between $\mathsf{DS}^E$ and $\mathsf{DS}^\varnothing$ any number of times. To summarize detection sequences that contain a certain pattern, we introduce the notion of a *detection term.*

**Definition 33 (Detection Term):** A *detection term* is a detection sequence or represents a (possibly infinite) set of detection sequences described using the following syntax:

$t_1|t_2$: The operator $|$ means an alternative, e.g., $t_1|t_2$ denotes that either the detection term $t_1$ occurs or the detection term $t_2$.

$\{\mathsf{t}\}$: The detection term $t$ occurs an arbitrary number of times, i.e., $\{\mathsf{t}\} = \epsilon|\mathsf{t}|\mathsf{t} \succ \mathsf{t}|\ldots$.

The operator $\succ$ may be used to link detection terms as well with the same semantical meaning. □

**Example 9:** Consider the development Enter $(\mathbf{O}, \mathbf{R})$. The detection sequences $\mathsf{DS}^E \succ \mathsf{DS}^\bullet \succ \mathsf{DS}^I$ as well as $\mathsf{DS}^E \succ \mathsf{DS}^\varnothing \succ \mathsf{DS}^I$ describe object trajectories that conform to Enter $(\mathbf{O}, \mathbf{R})$. Additionally, there exists an infinite number of detection sequences that conform to Enter $(\mathbf{O}, \mathbf{R})$ as well, like $\mathsf{DS}^E \succ \mathsf{DS}^\bullet \succ \mathsf{DS}^\varnothing \succ \mathsf{DS}^I$. The following detection term reflects this:

$$\mathsf{DS}^E \succ \left\{\mathsf{DS}^B|\mathsf{DS}^\bullet|\mathsf{DS}^\varnothing\right\} \succ \mathsf{DS}^I \tag{34}$$

**Definition 34 (Detection-Term Conformance):** *A detection sequence* $\mathbb{D}$ *conforms to a detection term* $t$ iff $\mathbb{D}$ contains a substring of detection scenarios that is represented by $t$. □

It is sufficient if a substring of a detection sequence conforms to the detection term because objects may move arbitrarily before or after conforming to the term.

**Example 10:** Continuing Example 9, suppose that object $\mathbf{O}$ crosses $\mathbf{R}$, i.e, $\mathbb{D}_{\mathbf{O}}^{\mathbf{R}} = \mathsf{DS}^E \succ \mathsf{DS}^\bullet \succ \mathsf{DS}^I \succ \mathsf{DS}^\bullet \succ \mathsf{DS}^E$. The substring $\mathsf{DS}^E \succ \mathsf{DS}^\bullet \succ \mathsf{DS}^I$ conforms to the detection term in (34) for Enter $(\mathbf{O}, \mathbf{R})$. ◆

There exist various algorithms, e.g., [32], to find a substring that conforms to a pattern. Section 7.4 provides detection terms similar to the one for Enter $(\mathbf{O}, \mathbf{R})$ above for every spatio-temporal development.

The detection term in (34) means that any $\mathbf{O}$ detected with $\mathsf{DS}^E$ at some time and later with $\mathsf{DS}^I$ conforms to Enter $(\mathbf{O}, \mathbf{R})$. It is not important which detection scenarios occur between $\mathsf{DS}^E$ and $\mathsf{DS}^I$ for $\mathbf{O}$ as long as the order described above is maintained. For a more concise presentation, we propose a relaxed version of the concatenation operator for detection scenarios:

**Definition 35 (Relaxed Detection Scenario Concatenation):** The relaxed concatenation of two detection scenarios $\mathsf{DS}_1 \tilde{\succ} \mathsf{DS}_2$ means that an object was detected according to $\mathsf{DS}_1$ at $t_1$ and later according to $\mathsf{DS}_2$ at $t_2$ with $t_1 < t_2$. □

**Lemma 17.** *Let* $\mathbb{DS} = \{\mathsf{DS}_1, \mathsf{DS}_2, \mathsf{DS}_3, \mathsf{DS}_4, \mathsf{DS}_5\}$ *be the domain of detection scenarios. If* $d = \mathsf{DS}_3|\mathsf{DS}_4|\mathsf{DS}_5$ *for* $\mathsf{DS}_i \neq \mathsf{DS}_j$ *with* $i \neq j$, *then* $\mathsf{DS}_1 \succ \{d\} \succ \mathsf{DS}_2 = \mathsf{DS}_1 \tilde{\succ} \mathsf{DS}_2$.

*Proof.* Follows directly from Lemma 3. ■

We illustrate the use of Lemma 17 by applying it to the detection term in (34): $\mathsf{DS}^E \succ \{\mathsf{DS}^B|\mathsf{DS}^\bullet|\mathsf{DS}^\varnothing\} \succ \mathsf{DS}^I$. In this case, $d = \mathsf{DS}^B|\mathsf{DS}^\bullet|\mathsf{DS}^\varnothing$, $\mathsf{DS}_1 = \mathsf{DS}^E$ and $\mathsf{DS}_2 = \mathsf{DS}^I$. Thus, we rewrite the term in (34) as $\mathsf{DS}^E \tilde{\succ} \mathsf{DS}^I$.

**Lemma 18.** $\mathsf{DS}_1 \succ \mathsf{DS}_2 \Rightarrow \mathsf{DS}_1 \tilde{\succ} \mathsf{DS}_2$

*Proof.* According to Definition 35, the right-hand side is true if $\mathsf{DS}_1$ occurs for some interval $[t_0, t_1[$ and $\mathsf{DS}_2$ occurs at $t_2 \geq t_1$. The left-hand side of the implication states that $\mathsf{DS}_1$ occurs for some interval $[t_0, t_1[$ and $\mathsf{DS}_2$ occurs at $t_2 = t_1$. Hence, if the left-hand side is true, the right-hand side is also true. ■

## *7.4 Detection Terms*

The inaccuracy of object detection or unobserved areas sometimes prevent a definite answer whether an object conforms to a given development or not. Given a development $\mathbb{P}$, the SN classifies objects detected into those that definitely conform ($\mathbb{P} = \mathcal{T}$), definitely do not conform ($\mathbb{P} = \mathcal{F}$) and maybe conform ($\mathbb{P} = \mathcal{M}$). In the following, we denote the true set of objects that conform to a development $\mathbb{P}$ with $\Omega_{\mathbb{P}}$.

**Definition 36 (Optimal Result):** The *result derived by the* SN *is optimal* iff a.) the set of objects where $\mathbb{P} = \mathcal{T}$ is a subset of $\Omega_{\mathbb{P}}$, b.) the set of objects where $\mathbb{P} = \mathcal{F}$ does not intersect with $\Omega_{\mathbb{P}}$ and c.) the set of objects where $\mathbb{P} = \mathcal{M}$ is minimal. □

In the following, we derive a *maximal* detection term for every element $\mathbb{P}$ in the canonical collection of developments.

**Definition 37 (Maximal Detection Term):** *detection term d is maximal* for a predicate sequence $\mathbb{P}$ iff it meets two conditions:

- There cannot exist an object $\mathbf{O}$ whose movement conforms to $\mathbb{P}$, but the corresponding detection sequence $\mathbb{D}_{\mathbf{O}}$ does not conform to $d$.
- There cannot exist an object $\mathbf{O}$ whose movement does not conform to $\mathbb{P}$, but the corresponding detection sequence $\mathbb{D}_{\mathbf{O}}$ conforms to $d$. □

**Example 11:** As we will show in Section 7.4.1, the term in (34) is maximal for Enter $(\mathbf{O},\mathbf{R})$. Contrary to (34), the following two terms are not maximal for Enter $(\mathbf{O},\mathbf{R})$:

$$\mathsf{DS}^E \succ \{\mathsf{DS}^B|\mathsf{DS}^\bullet\} \succ \mathsf{DS}^I \tag{35}$$

$$\mathsf{DS}^E \succ \{\mathsf{DS}^B|\mathsf{DS}^\bullet|\mathsf{DS}^\varnothing\} \tag{36}$$

The term in (35) is not maximal, because an object $\mathbf{O}$ detected with $\mathbb{D}_{\mathbf{O}}^{\mathbf{R}} = \mathsf{DS}^E \succ \mathsf{DS}^\bullet \succ \mathsf{DS}^\varnothing \succ \mathsf{DS}^I$ conforms to Enter $(\mathbf{O},\mathbf{R})$, but $\mathbb{D}_{\mathbf{O}}^{\mathbf{R}}$ does not conform to (35). Similarly, (36) is not maximal, because objects with $\mathbb{D}_{\mathbf{O}}^{\mathbf{R}} = \mathsf{DS}^E \succ \mathsf{DS}^\bullet \succ \mathsf{DS}^E$ do not conform to Enter $(\mathbf{O},\mathbf{R})$, but $\mathbb{D}_{\mathbf{O}}^{\mathbf{R}}$ conforms to (36). ◆

In the following, we address detection terms for developments in relation to regions and then those related to zones. For both types, we show that the derived result is optimal.

### 7.4.1 Detection Terms for Regions

Recall that the canonical collection of developments that describe the relationship of an object $\mathbf{O}$ and a region $\mathbf{R}$ has ten elements, listed in the left-hand column of Table 4. For each of these developments $\mathbb{P}(\mathbf{O},\mathbf{R})$, there is a detection term in the right-hand column such that $\mathbb{P}(\mathbf{O},\mathbf{R}) = \mathcal{T}$. We prove for each term that it is maximal in the context of the corresponding development $\mathbb{P}(\mathbf{O},\mathbf{R})$. Detection terms that indicate $\mathbb{P}(\mathbf{O},\mathbf{R}) = \mathcal{F}$ are addressed afterward.

#### Determining whether $\mathbb{P}(\mathbf{O},\mathbf{R}) = \mathcal{T}$.

The following Lemma is auxiliary, helping us to prove that the detection terms in Table 4 are maximal.

**Lemma 19.** *To ensure that Meet*$(\mathbf{O},\mathbf{R}) = \mathcal{T}$*, the detection sequence* $\mathbb{D}_{\mathbf{O}}^{\mathbf{R}}$ *of an object* $\mathbf{O}$ *must meet one of the following requirements:*

1. $\mathbb{D}_{\mathbf{O}}^{\mathbf{R}}$ *contains* $\mathsf{DS}^B$.
2. $\mathbb{D}_{\mathbf{O}}^{\mathbf{R}}$ *conforms to* $\mathsf{DS}^I \tilde{\succ} \mathsf{DS}^E$.
3. $\mathbb{D}_{\mathbf{O}}^{\mathbf{R}}$ *conforms to* $\mathsf{DS}^E \tilde{\succ} \mathsf{DS}^I$.

*For any other sequence, Meet*$(\mathbf{O},\mathbf{R})$ *yields* $\mathcal{M}$ *or* $\mathcal{F}$.

*Proof.* $\mathsf{DS}^B$ guarantees Meet $(\mathbf{O},\mathbf{R}) = \mathcal{T}$ according to (23). The other two cases imply that $\mathbf{O}$ has been detected on both sides of the border $\mathbf{R}^B$. Hence, between these detections there was a time when $\mathbf{O}$ was on $\mathbf{R}^B$ even if $\mathsf{DS}^B$

| $\mathbb{P}(\mathbf{O},\mathbf{R})$ | Detection term $d$ |
|---|---|
| Disjoint $(\mathbf{O},\mathbf{R}) \triangleright$ Meet $(\mathbf{O},\mathbf{R})$ | $\mathsf{DS}^E \succ \left\{\mathsf{DS}^\bullet \mid \mathsf{DS}^\varnothing\right\} \succ \left(\mathsf{DS}^B \mid \mathsf{DS}^I\right)$ |
| Inside $(\mathbf{O},\mathbf{R}) \triangleright$ Meet $(\mathbf{O},\mathbf{R})$ | $\mathsf{DS}^I \succ \left\{\mathsf{DS}^\bullet \mid \mathsf{DS}^\varnothing\right\} \succ \left(\mathsf{DS}^B \mid \mathsf{DS}^E\right)$ |
| Meet $(\mathbf{O},\mathbf{R}) \triangleright$ Disjoint $(\mathbf{O},\mathbf{R})$ | $\left(\mathsf{DS}^B \mid \mathsf{DS}^I\right) \succ \left\{\mathsf{DS}^\bullet \mid \mathsf{DS}^\varnothing\right\} \succ \mathsf{DS}^E$ |
| Meet $(\mathbf{O},\mathbf{R}) \triangleright$ Inside $(\mathbf{O},\mathbf{R})$ | $\left(\mathsf{DS}^B \mid \mathsf{DS}^E\right) \succ \left\{\mathsf{DS}^\bullet \mid \mathsf{DS}^\varnothing\right\} \succ \mathsf{DS}^I$ |
| Disjoint $(\mathbf{O},\mathbf{R}) \triangleright$ Meet $(\mathbf{O},\mathbf{R}) \triangleright$ Inside $(\mathbf{O},\mathbf{R})$ | $\mathsf{DS}^E \succ \left\{\mathsf{DS}^B \mid \mathsf{DS}^\bullet \mid \mathsf{DS}^\varnothing\right\} \succ \mathsf{DS}^I$ |
| Disjoint $(\mathbf{O},\mathbf{R}) \triangleright$ Meet $(\mathbf{O},\mathbf{R}) \triangleright$ Disjoint $(\mathbf{O},\mathbf{R})$ | $\mathsf{DS}^E \succ \mathsf{DS}^B \succ \mathsf{DS}^E$ |
| Inside $(\mathbf{O},\mathbf{R}) \triangleright$ Meet $(\mathbf{O},\mathbf{R}) \triangleright$ Disjoint $(\mathbf{O},\mathbf{R})$ | $\mathsf{DS}^I \succ \left\{\mathsf{DS}^B \mid \mathsf{DS}^\bullet \mid \mathsf{DS}^\varnothing\right\} \succ \mathsf{DS}^E$ |
| Inside $(\mathbf{O},\mathbf{R}) \triangleright$ Meet $(\mathbf{O},\mathbf{R}) \triangleright$ Inside $(\mathbf{O},\mathbf{R})$ | $\mathsf{DS}^I \succ \mathsf{DS}^B \succ \mathsf{DS}^I$ |
| Meet $(\mathbf{O},\mathbf{R}) \triangleright$ Disjoint $(\mathbf{O},\mathbf{R}) \triangleright$ Meet $(\mathbf{O},\mathbf{R})$ | $\left(\mathsf{DS}^I \mid \mathsf{DS}^B\right) \succ \left\{\mathsf{DS}^\bullet \mid \mathsf{DS}^\varnothing\right\} \succ \mathsf{DS}^E \succ \left\{\mathsf{DS}^E \mid \mathsf{DS}^\bullet \mid \mathsf{DS}^\varnothing\right\} \succ \left(\mathsf{DS}^I \mid \mathsf{DS}^B\right)$ |
| Meet $(\mathbf{O},\mathbf{R}) \triangleright$ Inside $(\mathbf{O},\mathbf{R}) \triangleright$ Meet $(\mathbf{O},\mathbf{R})$ | $\left(\mathsf{DS}^E \mid \mathsf{DS}^B\right) \succ \left\{\mathsf{DS}^\bullet \mid \mathsf{DS}^\varnothing\right\} \succ \mathsf{DS}^I \succ \left\{\mathsf{DS}^I \mid \mathsf{DS}^\bullet \mid \mathsf{DS}^\varnothing\right\} \succ \left(\mathsf{DS}^E \mid \mathsf{DS}^B\right)$ |

**Table 4** Detection terms which indicate $\mathbb{P}(\mathbf{O},\mathbf{R}) = \mathcal{T}$

did not occur. For instance, the object crossed the border while not being detected by any node. Detection sequences that do not meet either of these requirements conform to one of the two following terms:

- $\{\mathsf{DS}^E|\mathsf{DS}^\bullet|\mathsf{DS}^\varnothing\}$
- $\{\mathsf{DS}^I|\mathsf{DS}^\bullet|\mathsf{DS}^\varnothing\}$

Neither $\{\mathsf{DS}^E|\mathsf{DS}^\bullet|\mathsf{DS}^\varnothing\}$ nor $\{\mathsf{DS}^I|\mathsf{DS}^\bullet|\mathsf{DS}^\varnothing\}$ allow the SN to guarantee that $\mathrm{Meet}\,(\mathbf{O},\mathbf{R}) = \mathcal{T}$ according to (23). ■

Lemma 19 states that the SN can only guarantee $\mathrm{Meet}\,(\mathbf{O},\mathbf{R}) = \mathcal{T}$ if $\mathsf{DS}^B$ occurs, or if the object has been detected on both sides of the border. In any other case, $\mathrm{Meet}\,(\mathbf{O},\mathbf{R})$ yields $\mathcal{M}$ or $\mathcal{F}$.

**Lemma 20.** $\mathbb{P}\,(\mathbf{O},\mathbf{R}) = \mathcal{T}$ *iff* $\mathbb{D}_\mathbf{O}^\mathbf{R}$ *conforms to the corresponding detection term in Table 4.*

*Proof.* We prove this for every $\mathbb{P}\,(\mathbf{O},\mathbf{R})$ in the left-hand column of Table 4 separately: The movement of $\mathbf{O}$ conforms to $\mathrm{Disjoint}\,(\mathbf{O},\mathbf{R}) \triangleright \mathrm{Meet}\,(\mathbf{O},\mathbf{R})$ iff the detection sequence $\mathbb{D}_\mathbf{O}^\mathbf{R}$ conforms to $\mathsf{DS}^E \succ \{\mathsf{DS}^\bullet|\mathsf{DS}^\varnothing\} \succ (\mathsf{DS}^B|\mathsf{DS}^I)$. The reasoning for this is as follows: $\mathsf{DS}^E$ is the only detection scenario where $\mathrm{Disjoint}\,(\mathbf{O},\mathbf{R}) = \mathcal{T}$. According to Lemma 19, the SN can only guarantee $\mathrm{Meet}\,(\mathbf{O},\mathbf{R}) = \mathcal{T}$ after $\mathsf{DS}^E$ if $\mathsf{DS}^B$ or $\mathsf{DS}^I$ occurs. In the latter case, the detection term $\mathsf{DS}^E \tilde{\succ} \mathsf{DS}^I$ occurs. By applying Lemma 17, we rewrite this to $\mathsf{DS}^E \succ \{\mathsf{DS}^\bullet|\mathsf{DS}^B|\mathsf{DS}^\varnothing\} \succ \mathsf{DS}^I$. The only detection sequence not addressed by this term is $\mathsf{DS}^E \succ \mathsf{DS}^B$. Removing $\mathsf{DS}^B$ from $\{\mathsf{DS}^\bullet|\mathsf{DS}^B|\mathsf{DS}^\varnothing\}$ and adding it to the end of the detection term solves this. Hence, the resulting term is $\mathsf{DS}^E \succ \{\mathsf{DS}^\bullet|\mathsf{DS}^\varnothing\} \succ (\mathsf{DS}^B|\mathsf{DS}^I)$. The proof of correctness for detection terms related to all other developments consisting of two predicates is analogous.

According to Lemma 19, to derive that $\mathrm{Enter}\,(\mathbf{O},\mathbf{R}) = \mathcal{T}$ or $\mathrm{Leave}\,(\mathbf{O},\mathbf{R}) = \mathcal{T}$, $\mathbf{O}$ must be detected conforming to $\mathsf{DS}^E \tilde{\succ} \mathsf{DS}^I$ and $\mathsf{DS}^I \tilde{\succ} \mathsf{DS}^E$ respectively. By applying Lemma 17, both terms are rewritten to the corresponding detection terms in Table 4.

For $\mathrm{Disjoint}\,(\mathbf{O},\mathbf{R}) \triangleright \mathrm{Meet}\,(\mathbf{O},\mathbf{R}) \triangleright \mathrm{Disjoint}\,(\mathbf{O},\mathbf{R})$, the SN must detect $\mathbf{O}$ with $\mathsf{DS}^E$ first immediately followed by $\mathsf{DS}^B$ and $\mathsf{DS}^E$. If either $\mathsf{DS}^\bullet$ or $\mathsf{DS}^\varnothing$ occur in between, $\mathbf{O}$ could have moved into $\mathbf{R}$ for some time. Such a movement would not conform to $\mathrm{Disjoint}\,(\mathbf{O},\mathbf{R}) \triangleright \mathrm{Meet}\,(\mathbf{O},\mathbf{R}) \triangleright \mathrm{Disjoint}\,(\mathbf{O},\mathbf{R})$. Thus, the term in Table 4 is correct. The proof for the detection term of $\mathrm{Inside}\,(\mathbf{O},\mathbf{R}) \triangleright \mathrm{Meet}\,(\mathbf{O},\mathbf{R}) \triangleright \mathrm{Inside}\,(\mathbf{O},\mathbf{R})$ is analogous.

We consider $\mathrm{Meet}\,(\mathbf{O},\mathbf{R}) \triangleright \mathrm{Disjoint}\,(\mathbf{O},\mathbf{R})$ in the development $\mathrm{Meet}\,(\mathbf{O},\mathbf{R}) \triangleright \mathrm{Disjoint}\,(\mathbf{O},\mathbf{R}) \triangleright \mathrm{Meet}\,(\mathbf{O},\mathbf{R})$ first: To conform to this first part, the object $\mathbf{O}$ must be detected with $\mathsf{DS}^I \tilde{\succ} \mathsf{DS}^E$ or $\mathsf{DS}^B \succ \mathsf{DS}^E$ (cf. Lemma 19). Hence, $(\mathsf{DS}^I|\mathsf{DS}^B) \succ \{\mathsf{DS}^\bullet|\mathsf{DS}^\varnothing\} \succ \mathsf{DS}^E$ guarantees the first part, i.e., $\mathrm{Meet}\,(\mathbf{O},\mathbf{R}) \triangleright \mathrm{Disjoint}\,(\mathbf{O},\mathbf{R})$. Similarly, to conform to $\mathrm{Disjoint}\,(\mathbf{O},\mathbf{R}) \triangleright \mathrm{Meet}\,(\mathbf{O},\mathbf{R})$, the object $\mathbf{O}$ must be detected with $\mathsf{DS}^E \tilde{\succ} \mathsf{DS}^I$ or $\mathsf{DS}^E \succ \mathsf{DS}^B$. Rewriting this by applying Lemma 17 yields the corresponding detection term in Table 4. The proof for the detection term for $\mathrm{Meet}\,(\mathbf{O},\mathbf{R}) \triangleright \mathrm{Inside}\,(\mathbf{O},\mathbf{R}) \triangleright \mathrm{Meet}\,(\mathbf{O},\mathbf{R})$ is analogous. ■

Summing up, we have shown how SN can derive $\mathbb{P}(\mathbf{O},\mathbf{R}) = \mathcal{T}$ by providing a detection term for every spatio-temporal development.

**Determining whether $\mathbb{P}(\mathbf{O},\mathbf{R}) = \mathcal{F}$.**

Now we show how SN derive $\mathbb{P}(\mathbf{O},\mathbf{R}) = \mathcal{F}$. The most important difference to $\mathbb{P}(\mathbf{O},\mathbf{R}) = \mathcal{T}$ is that one must consider the whole detection sequence instead of a substring: While it is sufficient to find a substring in the detection sequence that conforms to a detection term to determine that $\mathbb{P}(\mathbf{O},\mathbf{R}) = \mathcal{T}$, to compute $\mathbb{P}(\mathbf{O},\mathbf{R}) = \mathcal{F}$ the SN must rule out that any part of the detection sequence could conform to $\mathbb{P}(\mathbf{O},\mathbf{R})$.

**Lemma 21.** *An object* $\mathbf{O}$ *which is detected according to* $\mathsf{DS}^{\bullet}$ *could conform to any spatio-temporal development* $\mathbb{P}(\mathbf{O},\mathbf{R})$.

*Proof.* According to Definition 23, $\mathsf{DS}^{\bullet}$ means that $\mathbf{POP}^{\mathbf{O}}_{t}$ intersects with all partitions of $\mathbf{R}$. This means that the position of $\mathbf{O}$ is so "close" to the border that the sensor network cannot provide a definite answer on which side of the border $\mathbf{O}$ is. Thus, an object could repeatedly move around and over the border of $\mathbf{R}$ in any way while the sensor network can only determine $\mathsf{DS}^{\bullet}$. During this time, $\mathbf{O}$ could fulfill any development that describes the relationship between $\mathbf{O}$ and $\mathbf{R}$. ■

Lemma 21 implies that detection sequences that do not conform to a development must not contain $\mathsf{DS}^{\bullet}$. Looking at Table 2, this also applies to $\mathsf{DS}^{\varnothing}$. Typically detection areas may have any shape or size, i.e., objects can cross the border of a region in arbitrary ways while being undetected. This changes if assumptions about the space covered by detection areas are viable, e.g., for controlled deployments. We discuss three such *coverage assumptions (CA)* in the following:

**No assumption ($\mathbf{CA}^{\varnothing}$)**: We assume that nodes have been deployed randomly, and it is not fixed a priori which parts of space are observed.

**Coverage Assumption Border ($\mathbf{CA}^{B}$)**: Nodes have been deployed in such a way that their detection areas cover the border $\mathbf{R}^{B}$ entirely.

**Coverage Assumption Border Interior ($\mathbf{CA}^{BI}$)**: The deployment guarantees that objects inside as well as objects on the border are detected. Thus, $\mathsf{DS}^{\varnothing}$ only occurs for objects that are in $\mathbf{R}^{E}$.

**Lemma 22.** *In case of* $CA^{\varnothing}$, *an object* $\mathbf{O}$ *that is temporarily undetected, i.e.,* $\mathsf{DS}^{\varnothing}$ *occurs at least once in* $\mathbb{D}^{\mathbf{R}}_{\mathbf{O}}$, *could conform to any development* $\mathbb{P}(\mathbf{O},\mathbf{R})$.

*Proof.* As stated above, detection areas may have any size or shape and thus the set of points that is unobserved could intersect with any partition of $\mathbf{R}$. An undetected object $\mathbf{O}$ could be at any of these unobserved points in space and thus in any partition of **R**. Hence, **O** may conform to any development that describes the relation between **O** and **R**. ■

According to Lemma 22, any occurrence of $\mathsf{DS}^{\varnothing}$ or $\mathsf{DS}^{\bullet}$ in the detection sequence rules out $\mathbb{P}(\mathbf{O},\mathbf{R}) = \mathcal{F}$ if assumptions about the coverage of space are not viable. SN with $\mathrm{CA}^{\varnothing}$ can only derive $\mathbb{P}(\mathbf{O},\mathbf{R}) = \mathcal{F}$ if the object is detected according to either $\mathsf{DS}^{I}$, $\mathsf{DS}^{B}$ or $\mathsf{DS}^{E}$ at all times. Hence, $\mathbb{D}_{\mathbf{O}}^{\mathbf{R}}$ must equal $\{\mathsf{DS}^{I}|\mathsf{DS}^{B}|\mathsf{DS}^{E}\}$ as shown in Equation (37).

For SN with $\mathrm{CA}^{B}$, we can assume that objects do not cross the border while being undetected. To derive that $\mathbb{P}(\mathbf{O},\mathbf{R}) = \mathcal{F}$, the SN must ensure first that the detection sequence of **O** does not conform to the corresponding detection term in Table 4. Once this condition is met, it is certain that $\mathbb{P}(\mathbf{O},\mathbf{R}) = \mathcal{F}$ if the detection sequence $\mathbb{D}_{\mathbf{O}}^{\mathbf{R}}$ does not contain $\mathsf{DS}^{\bullet}$ (cf. Equation (38)).

The reasoning for $\mathrm{CA}^{B}$ applies to SN with $\mathrm{CA}^{BI}$ as well. Additionally, any undetected object must be outside of the region **R**, i.e., in $\mathbf{R}^{E}$. Thus, we replace any occurrence of $\mathsf{DS}^{\varnothing}$ with $\mathsf{DS}^{E}$ prior to determining if the detection sequence of **O** conforms to the term in Table 4 associated with $\mathbb{P}(\mathbf{O},\mathbf{R})$.

**Summary – Development results for queries with regions.**

Given a detection term $d$ associated with a development $\mathbb{P}(\mathbf{O},\mathbf{R})$, Equations 37-39 summarize our findings regarding the translation of sequences of object detections into the result of a development $\mathbb{P}(\mathbf{O},\mathbf{R})$.

$$\mathbb{P}_{\mathrm{CA}^{\varnothing}}(\mathbf{O},\mathbf{R}) = \begin{cases} \mathcal{T} & \text{iff } \mathbb{D}_{\mathbf{O}}^{\mathbf{R}} \text{ conforms to the corresponding detection term } d \text{ in Table 4} \\ \mathcal{F} & \text{iff } \mathbb{D}_{\mathbf{O}}^{\mathbf{R}} \text{ does not conform to } d \text{ and } \mathbb{D}_{\mathbf{O}}^{\mathbf{R}} = \{\mathsf{DS}^{I}|\mathsf{DS}^{B}|\mathsf{DS}^{E}\} \\ \mathcal{M} & \text{Otherwise} \end{cases} \tag{37}$$

$$\mathbb{P}_{\mathrm{CA}^{B}}(\mathbf{O},\mathbf{R}) = \begin{cases} \mathcal{T} & \text{iff } \mathbb{D}_{\mathbf{O}}^{\mathbf{R}} \text{ conforms to the corresponding detection term } d \text{ in Table 4} \\ \mathcal{F} & \text{iff } \mathbb{D}_{\mathbf{O}}^{\mathbf{R}} \text{ does not conform to } d \text{ and } \mathbb{D}_{\mathbf{O}}^{\mathbf{R}} = \{\mathsf{DS}^{I}|\mathsf{DS}^{E}|\mathsf{DS}^{\varnothing}|\mathsf{DS}^{B}\} \\ \mathcal{M} & \text{Otherwise} \end{cases} \tag{38}$$

$$\mathbb{P}_{\mathrm{CA}^{BI}}(\mathbf{O},\mathbf{R}) = \begin{cases} \mathcal{T} & \text{iff } \mathbb{D}_{\mathbf{O}}^{\mathbf{R}} \text{ conforms to the corresponding detection term } d \\ & \text{in Table 4 with } \mathsf{DS}^{\varnothing} \text{ replaced by } \mathsf{DS}^{E} \\ \mathcal{F} & \text{iff } \mathbb{D}_{\mathbf{O}}^{\mathbf{R}} \text{ does not conform to } d \text{ and } \mathbb{D}_{\mathbf{O}}^{\mathbf{R}} = \{\mathsf{DS}^{I}|\mathsf{DS}^{E}|\mathsf{DS}^{B}\} \\ \mathcal{M} & \text{Otherwise} \end{cases} \tag{39}$$

**Theorem 1.** *The results for developments that describe the relationship of an object and region derived according to Equations 37-39 are optimal.*

*Proof.* Let $\Omega_{\mathbb{P}(\mathbf{O},\mathbf{R})}$ be the set of objects that conform to a development $\mathbb{P}(\mathbf{O},\mathbf{R})$ in question. The set of objects where $\mathbb{P}(\mathbf{O},\mathbf{R}) = \mathcal{T}$ is the largest subset of $\Omega_{\mathbb{P}(\mathbf{O},\mathbf{R})}$ a sensor network can derive according to the lemmas in Section 7.4.1. Similarly, the set of objects where $\mathbb{P}(\mathbf{O},\mathbf{R}) = \mathcal{F}$ is the largest superset of $\Omega_{\mathbb{P}(\mathbf{O},\mathbf{R})}$ the sensor network can derive. Therefore, the set of objects where $\mathbb{P}(\mathbf{O},\mathbf{R}) = \mathcal{M}$ is minimal, i.e., contains only objects where the accuracy of the object detection prevents a definitive answer. ■

### 7.4.2 Detection Terms for Zones

According to Table 3, all predicates that express the relationship between an object and a zone yield $\mathcal{T}$ or $\mathcal{F}$, but never $\mathcal{M}$. Furthermore, the table shows that for any predicate $\mathrm{P}(\mathbf{O},\mathrm{Z})$, there exists exactly one detection scenario $\mathsf{DS}_i$ which yields $\mathrm{P}(\mathbf{O},\mathrm{Z}) = \mathcal{T}$. All other detection scenarios $\mathsf{DS}_j \neq \mathsf{DS}_i$ yield $\mathrm{P}(\mathbf{O},\mathrm{Z}) = \mathcal{F}$. Compared to regions, this eases the translation of detection sequences to development results considerably.

**Lemma 23.** *Let* $\mathsf{DS}_i$ *be the detection scenario which yields* $P_i(\mathbf{O},\mathrm{Z}) = \mathcal{T}$, *and* $\mathsf{DS}_j$ *is the detection scenario which yields* $P_j(\mathbf{O},\mathrm{Z}) = \mathcal{T}$. *If the detection sequence* $\mathbb{D}^{\mathrm{Z}}_{\mathbf{O}}$ *conforms to the term* $\mathsf{DS}_i \succ \mathsf{DS}_j$ *(cf. Definition 34), then* $P_i(\mathbf{O},\mathrm{Z}) \triangleright P_j(\mathbf{O},\mathrm{Z}) = \mathcal{T}$. *If* $\mathbb{D}^{\mathrm{Z}}_{\mathbf{O}}$ *does not conform to* $\mathsf{DS}_i \succ \mathsf{DS}_j$, *then* $P_i(\mathbf{O},\mathrm{Z}) \triangleright P_j(\mathbf{O},\mathrm{Z}) = \mathcal{F}$.

*Proof.* We prove $\mathrm{P}_i(\mathbf{O},\mathrm{Z}) \triangleright \mathrm{P}_j(\mathbf{O},\mathrm{Z}) = \mathcal{T}$ first: According to Definitions 31 and 34, conformance of $\mathbb{D}^{\mathrm{Z}}_{\mathbf{O}}$ to $\mathsf{DS}_i \succ \mathsf{DS}_j$ means that the object $\mathbf{O}$ was detected with $\mathsf{DS}_i$ during $[t_1, t_2[$ and then with $\mathsf{DS}_j$ at $t_2$. Since $\mathsf{DS}_i$ yields $\mathrm{P}_i(\mathbf{O},\mathrm{Z}) = \mathcal{T}$, we derive that $\mathrm{P}_i(\mathbf{O},\mathrm{Z}) = \mathcal{T}$ for the interval $[t_1, t_2[$ and $\mathrm{P}_i(\mathbf{O},\mathrm{Z}) = \mathcal{T}$ at $t_2$. Hence, $\mathrm{P}_i(\mathbf{O},\mathrm{Z}) \triangleright \mathrm{P}_j(\mathbf{O},\mathrm{Z}) = \mathcal{T}$.

If $\mathbb{D}^{\mathrm{Z}}_{\mathbf{O}}$ does not conform to $\mathsf{DS}_i \succ \mathsf{DS}_j$, there is no substring in $\mathbb{D}^{\mathrm{Z}}_{\mathbf{O}}$ where $\mathsf{DS}_i$ is followed by $\mathsf{DS}_j$. This means that either $\mathsf{DS}_j$ never follows $\mathsf{DS}_i$, or $\mathsf{DS}_i$ or $\mathsf{DS}_j$ never occur. For all of these cases, the sensor network can guarantee that $\mathbf{O}$ does not fulfill $\mathrm{P}_i(\mathbf{O},\mathrm{Z}) \triangleright \mathrm{P}_j(\mathbf{O},\mathrm{Z})$ and thus return $\mathcal{F}$. ■

**Lemma 24.** *Let* $\mathsf{DS}_i$ *be the detection scenario which yields* $P_i(\mathbf{O},\mathrm{Z}) = \mathcal{T}$, *and* $\mathsf{DS}_j$ *is the detection scenario which yields* $P_j(\mathbf{O},\mathrm{Z}) = \mathcal{T}$. $P_i(\mathbf{O},\mathrm{Z}) \mathbin{\widetilde{\triangleright}} P_j(\mathbf{O},\mathrm{Z}) = \mathcal{T}$ *if* $\mathbb{D}^{\mathrm{Z}}_{\mathbf{O}}$ *of* $\mathbf{O}$ *conforms to* $\mathsf{DS}_i \mathbin{\widetilde{\succ}} \mathsf{DS}_j$

*Proof.* Analogous to Lemma 23. ■

Lemmas 23 and 24 ease the definition of detection terms for any of the 584 developments with zones. Due to the large number of developments with zones, we do not list a detection term for each one in this paper and explain how to derive maximal detection terms based on these lemmas: Consider a development $\mathbb{P}(\mathbf{O},\mathrm{Z}) = \mathrm{P}_1(\mathbf{O},\mathrm{Z})\,\theta_1 \mathrm{P}_2(\mathbf{O},\mathrm{Z})\,\theta_2 \ldots \theta_{q-1}\mathrm{P}_q(\mathbf{O},\mathrm{Z})$ where $\theta_i$ represents any concatenation operator, i.e., $\theta_i \in \{\triangleright, \widetilde{\triangleright}\}$. Let $\mathsf{DS}_i$ be the detection scenario where $\mathrm{P}_i(\mathbf{O},\mathrm{Z}) = \mathcal{T}$ according to Table 3. Thus, the detection term starts with $\mathsf{DS}_1$ and the second detection scenario in the term is $\mathsf{DS}_2$. If the concatenation operator between $\mathrm{P}_1(\mathbf{O},\mathrm{Z})$ and $\mathrm{P}_2(\mathbf{O},\mathrm{Z})$ is $\triangleright$, then the detection term starts with $\mathsf{DS}_1 \succ \mathsf{DS}_2$. Otherwise, the detection terms starts with $\mathsf{DS}_1 \tilde{\succ} \mathsf{DS}_2$. Next, we consider $\mathrm{P}_3(\mathbf{O},\mathsf{Z})$ and how it is concatenated to $\mathrm{P}_2(\mathbf{O},\mathsf{Z})$. This continues until a detection scenario corresponding to $\mathrm{P}_q(\mathbf{O},\mathsf{Z})$ terminates the detection term. For example, Enter $(\mathbf{O},\mathsf{Z})$ defined in (29) has the detection term $\mathsf{DS}^E \succ \mathsf{DS}^B \succ \mathsf{DS}^I$.

**Theorem 2.** *Suppose $\Omega_{\mathbb{P}(\mathbf{O},\mathsf{Z})}$ is the set of objects that conform to a development $\mathbb{P}(\mathbf{O},\mathsf{Z})$. The set of objects determined by the* SN *where $\mathbb{P}(\mathbf{O},\mathsf{Z}) = \mathcal{T}$ equals $\Omega_{\mathbb{P}(\mathbf{O},\mathsf{Z})}$.*

*Proof.* Directly follows from Lemmas 23 and 24 and the fact that there does not exist a predicate $\mathrm{P}(\mathbf{O},\mathsf{Z})$ which yields $\mathcal{M}$ for any detection scenario. ■

This concludes our discussion regarding the contributions *Semantics* and *Optimality*. The remainder of this paper addresses the contribution *Efficiency*.

## 8 Spatio-Temporal Query Processing in SN

We have implemented a distributed query processor for spatio-temporal queries in SN. This section outlines the core mechanisms of the query processor as follows: First, Section 8.1 proposes a set of data structures used for the computation of detection scenarios (Section 8.2). Second, we describe how to to collect the information required for this computation at the base station (Section 8.3). The core contribution of this section is a proposal how to process spatio-temporal queries. This includes two execution strategies that reduce the number of messages required to process such queries (Section 8.4). Reducing the number of messages is important, since energy consumption due to communication typically dominates the overall energy consumption of sensor nodes [1, 41]. Mechanisms that aim at regions or dynamic zones are omitted here due to lack of space.

The query processor must return every object $\mathbf{O}$ that conforms to $\mathbb{P}(\mathbf{O},\mathsf{Z})$. Prior to processing a query, the following steps must be completed:

1. Definition of a zone $\mathsf{Z}$.
2. Specification of the movement of interest as a spatio-temporal development $\mathbb{P}(\mathbf{O},\mathsf{Z})$.
3. Dissemination of a list of nodes representing $\mathsf{Z}$ and the query $\mathbb{P}(\mathbf{O},\mathsf{Z})$ to all nodes.

The SN must compute the detection scenario whenever an object is detected. Based on the detection scenario, the SN determines if a predicate of the query is true using Table 3. Thus, it is sufficient to limit the discussion in the following to deriving detection scenarios from object detections.

The distributed strategies notify the base station whenever a predicate $\mathrm{P}(\mathbf{O},\mathsf{Z})$ in $\mathbb{P}(\mathbf{O},\mathsf{Z})$ is satisfied. The base station determines if $\mathbf{O}$ has fulfilled a development $\mathbb{P}(\mathbf{O},\mathsf{Z})$ using these notifications. Note that a node may send several notifications regarding a predicate to the base station because it detects the same object more than once. This is intended, for two reasons: First, the query $\mathbb{P}(\mathbf{O}, \mathsf{Z})$ may contain a single predicate more than once, e.g., Touch$(\mathbf{O}, \mathsf{Z})$. Second, coordinating nodes to prevent such redundant notifications requires communication. A pre-study of ours has shown that such a coordination only pays off if the network is very small, the zone is small, and if the object moves through detection areas of most nodes repeatedly. Thus, we do not intend to prevent this. On the other hand, we show in Section 8.4 how to exploit spatio-temporal semantics to reduce the number of notifications.

## *8.1 Data Structures and Algorithms*

To store the information on objects detected, we use a list Detections. It depends on the strategy where Detections is stored: For the centralized strategy, it is at the base station. Contrary to that, the distributed strategies share and replicate the elements of Detections in such a way that sensor nodes can compute detection scenarios based on it. Every element of Detections represents the detection of an object $\mathbf{O}$ by a node $\mathcal{S}_i$ during a time interval $[t_{entry}, t_{exit}]$. Thus, every element of Detections has the following structure:

`NodeID`: Identifier of the node $\mathcal{S}_i$ detecting $\mathbf{O}$.

`ObjectID`: An identifier of the object $\mathbf{O}$ that has been detected by $\mathcal{S}_i$.

$t^{\mathbf{O}}_{entry}$: The time at which $\mathbf{O}$ has entered the detection area of $\mathcal{S}_i$.

$t^{\mathbf{O}}_{exit}$: This value either equals $\emptyset$ or a time $t > t_{entry}$. If it equals $\emptyset$, this indicates that $\mathcal{S}_i$ is still detecting $\mathbf{O}$. Otherwise, this value equals the time $t_{exit}$ at which $\mathbf{O}$ has left the detection area of $\mathcal{S}_i$.

We say *an element* E *of Detections originates from node* $\mathcal{S}_i$ *if* E.NodeID $= \mathcal{S}_i$. When an object $\mathbf{O}$ enters the detection area $\mathbf{DA}_i$ at $t_1$, the corresponding node $\mathcal{S}_i$ generates a list entry $[\mathcal{S}_i, \mathbf{O}, t_1, \emptyset]$ and stores it in Detections. Once $\mathbf{O}$ leaves $\mathbf{DA}_i$ at $t_2$, this list entry is updated to $[\mathcal{S}_i, \mathbf{O}, t_1, t_2]$. Note that an object that repeatedly enters and leaves the detection area of a node may result in several list elements originating from the same node.

For non-continuous detection mechanisms nodes can determine $t_{entry}$ and $t_{exit}$ by temporal interpolation: Suppose $\mathcal{S}_i$ checks periodically at $t_0, t_1, \ldots$ for objects. An entry occurs at $t_j$ if $\mathcal{S}_i$ did not detect an object at $t_{j-1}$ but detects it at $t_j$, i.e., $t_{entry} = t_j$. An exit occurs at $t_j$ if $\mathcal{S}_i$ detected an object at $t_j$ and does not detect it at $t_{j+1}$, i.e., $t_{exit} = t_{j+1}$. Research on detection mechanisms reviewed in Section 3.1 has yielded approaches to detect continuously moving objects using non-continuous detection mechanisms with limited temporal resolution.

## 8.2 Computing Detection Scenarios

According to Section 6.3, the sensor network must compute how the detection set $\mathsf{DetSet}_t^{\mathbf{O}}$ intersects with $\mathsf{Z}$ and $\overline{\mathsf{Z}}$, to compute a detection scenario at time $t$ for a given object $\mathbf{O}$. We refer to this computation as $isDetecting\,(\mathsf{S}^*, t, \mathbf{O})$. Its result is as follows:

$$isDetecting\,(\mathsf{S}^*, t, \mathbf{O}) = \begin{cases} \mathcal{T} \text{ iff } \exists \mathcal{S}_i \in \mathsf{S}^* : detect\,(\mathcal{S}_i, \mathbf{O}, t) \\ \mathcal{F} \text{ Otherwise} \end{cases}$$

The input parameter $\mathsf{S}^*$ is either $\mathsf{Z}$ or $\overline{\mathsf{Z}}$. The implementation of the function $isDetecting\,(\mathsf{S}^*, t, \mathbf{O})$ is straightforward: It consists of a single iteration over $\mathsf{Detections}$ and $\mathsf{Z}$. Due to this simplicity, an algorithm is omitted here. By computing $isDetecting\,(\mathsf{Z}, t, \mathbf{O})$ and then $isDetecting\,(\overline{\mathsf{Z}}, t, \mathbf{O})$, we obtain two boolean values which indicate whether $\mathsf{Z}$ and $\overline{\mathsf{Z}}$ intersect with $\mathsf{DetSet}_t^{\mathbf{O}}$. According to Lemma 8, this is sufficient to compute a detection scenario in the context of a zone, and Table 5 illustrates this: Each cell corresponds to a pair of booleans that represent the result of the calls to $isDetecting\,(\mathsf{Z}, t, \mathbf{O})$ and $isDetecting\,(\overline{\mathsf{Z}}, t, \mathbf{O})$ and contains the corresponding detection scenario.

| | | $isDetecting\,(\overline{\mathsf{Z}}, t, \mathbf{O})$ $\mathcal{T}$ | $\mathcal{F}$ |
|---|---|---|---|
| $isDetecting\,(\mathsf{Z}, t, \mathbf{O})$ | $\mathcal{T}$ | $\mathsf{DS}^B$ | $\mathsf{DS}^I$ |
| | $\mathcal{F}$ | $\mathsf{DS}^E$ | $\mathsf{DS}^{\varnothing}$ |

**Table 5** Deriving detection scenarios using $isDetecting\,(\mathsf{S}^*, t, \mathbf{O})$ with $\mathsf{S}^* \in \{\mathsf{Z}, \overline{\mathsf{Z}}\}$

We now address the collection of the elements in $\mathsf{Detections}$ to ensure that the result of the detection-scenario computation according to Table 5 is correct.

**Definition 38 (Correctness):** The *computation of a detection scenario* $\mathsf{DS}^*$ *is correct* if the space partition that corresponds to $\mathsf{DS}^*$ (cf. Definitions 24-26) contains the position $\mathsf{p} \in \mathbb{E}^d$ of the object detected. □

**Definition 39 (Completeness):** The *list* $\mathsf{Detections}$ *is complete regarding an object* $\mathbf{O}$ *and a time* $t$ if $\mathsf{Detections}$ contains all existing elements $[\mathcal{S}_i, \mathbf{O}, t_1, t_2]$ with $t_1 \leq t$ and $t \leq t_2$ or $t_2 = \emptyset$. □

**Lemma 25.** *If* $\mathsf{Detections}$ *is complete, the detection scenario computed according to Table 5 is correct.*

*Proof.* Without loss of generality, assume the computed detection scenario regarding an object $\mathbf{O}$ and a time $t$ is $\mathsf{DS}^E$. According to Definition 16, this means that $\mathbf{O}$ is in $\mathsf{Z}^E$. Considering Lemma 8, this implies that there is at least one node $\mathcal{S}_i \in \overline{\mathsf{Z}}$ that detects $\mathbf{O}$. The computed detection scenario would be incorrect, if there existed another node $\mathcal{S}_j \in \mathsf{Z}$ which detects $\mathbf{O}$ at $t$ as well. Such a node cannot exist since Detections and Zones are complete. For the other detection scenarios, the proof is similar. ∎

Summing up, the base station or an arbitrary sensor node must store a complete list Detections to compute a detection scenario for a given object $\mathbf{O}$ and a time $t$. Our goal in the following is acquiring a complete list Detections while minimizing the number of messages.

## *8.3 Centralized data collection*

Notifying the base station whenever an object enters or leaves the detection area of an arbitrary node $\mathcal{S}_i$ is a straightforward approach to achieve completeness. For every incoming notification, the base station can modify its version of Detections and compute a detection scenario, as shown in Algorithm 8.1.

The first part of Algorithm 8.1 is executed by any node $\mathcal{S}_i$ detecting objects and results in a notification for every object detection. If $\mathcal{S}_i$ is not a communication neighbor of the base station, the notification is forwarded to the base station via multiple hops. The base station executes the second part once it receives the notification and modifies Detections accordingly. Prior to computing the detection scenario, the base station has to wait $t_{delay}$. This ensures that notifications of nodes which simultaneously detect an object have arrived before the detection scenario is computed. $t_{delay}$ is the maximum time a notification may need to be forwarded to the base station. Its actual value depends on factors such as communication hardware, SN size, routing protocol etc. For our reference implementation we use a delay of 30 seconds.

**Algorithm 8.1:** Centralized data collection

**When** $\mathbf{O}$ *enters/leaves* $\mathbf{DA}_i$ *of* $\mathcal{S}_i$ *at* $t$ **do**
  $\mathcal{S}_i$ sends corresponding notification to base station
**end**
**When** *base station receives notification from* $\mathcal{S}_i$ **do**
  Modify `Detections` at base station
  Wait $t_{delay}$
  Compute $[\mathit{isDetecting}(\mathsf{Z}, t, \mathbf{O}), \mathit{isDetecting}(\overline{\mathsf{Z}}, t, \mathbf{O})]$
**end**

## 8.4 Distributed data collection

In the following, we show how to distribute Detections in a way that allows nodes to compute detection scenarios while only storing a part of Detections. This reduces communication for two reasons:

- Nodes only notify the base station of objects that at least fulfill one $\mathrm{P}(\mathbf{O}, \mathsf{Z}) \in \mathbb{P}(\mathbf{O}, \mathsf{Z})$.
- To compute detection scenarios, nodes only communicate with nodes in their vicinity, i.e., multi-hop messages only occur if an object fulfills a predicate of the query.

The latter point stems from the following idea: When a node $\mathcal{S}_i$ detects an object $\mathbf{O}$, only nodes in its vicinity can detect the object at the same time. This is because $\mathbf{O}$ at position $\mathsf{p} \in \mathbb{E}^d$ can be detected only by nodes whose detection area contains $\mathsf{p}$. Even though $\mathsf{p}$ is typically unknown due to the inaccuracy of the detection mechanism, one can derive that only nodes close to $\mathcal{S}_i$ could possibly detect $\mathbf{O}$ at the same time.

**Definition 40 (Detection Neighbor):** Node $\mathcal{S}_j$ is a *detection neighbor of* $\mathcal{S}_i$ if the detection areas of both nodes overlap, i.e., $\mathbf{DA}_i \cap \mathbf{DA}_j \neq \varnothing$. $\mathsf{DetNeigh}_i$ is the set of detection neighbors of $\mathcal{S}_i$. □

As discussed in Section 3.1, detection areas are indeterminable for some SN. We show in Section 8.4.1 how a node can approximate its detection neighbors.

**Notation (Detection Neighbor Subsets):** The subset of detection neighbors of a node $\mathcal{S}_i$ that are in $\mathsf{Z}$ are denoted by $\mathsf{DetNeigh}_i^{\mathsf{Z}}$. Similarly, $\mathsf{DetNeigh}_i^{\overline{\mathsf{Z}}}$ contains all detection neighbors of $\mathcal{S}_i$ that are outside of $\mathsf{Z}$.

Every node $\mathcal{S}_i$ can derive for each detection neighbor $\mathcal{S}_j \in \mathsf{DetNeigh}_i$ if it is in $\mathsf{Z}$ or not since the query has been disseminated to all nodes previously.

**Lemma 26.** *Detections stored at $\mathcal{S}_i$ is complete regarding the object $\mathbf{O}$ and time $t$ if $\mathcal{S}_i$ detects $\mathbf{O}$ at $t$ and obtains all list elements for Detections regarding $\mathbf{O}$ originating from its detection neighbors* $\mathsf{DetNeigh}_i$.

*Proof.* We prove this by showing that there cannot exist a node $\mathcal{S}_j \notin \mathsf{DetNeigh}_i$ that detects $\mathbf{O}$ at $t$. $\mathcal{S}_j \notin \mathsf{DetNeigh}_i$ implies that the detection areas of $\mathcal{S}_i$ and $\mathcal{S}_j$ do not overlap, i.e., $\mathbf{DA}_i \cap \mathbf{DA}_j = \varnothing$. Thus, there does not exist a $\mathsf{p} \in \mathbb{E}^d$ where $\mathcal{S}_i$ and $\mathcal{S}_j$ can detect $\mathbf{O}$ simultaneously. Hence, $\mathcal{S}_j$ cannot detect $\mathbf{O}$ at $t$. ■

Lemma 26 limits the nodes from which $\mathcal{S}_i$ must acquire list elements for Detections to the detection neighbors $\mathsf{DetNeigh}_i$. By taking into account that $\mathcal{S}_i$ is either in $\mathsf{Z}$ or $\overline{\mathsf{Z}}$, we actually can compute a correct detection scenario without Detections being complete.

**Definition 41 (Semi-Completeness):** *Detections regarding $\mathbf{O}$ and $t$ stored at a node $\mathcal{S}_i \in \mathsf{Z}$ is semi-complete* if it contains all list elements $[\mathcal{S}_j, \mathbf{O}, t_1, t_2]$ with $t_1 \leq t \leq t_2$ where $\mathcal{S}_j \in \mathsf{DetNeigh}_i^{\overline{\mathsf{Z}}}$. Analogously, *Detections regarding* $\mathbf{O}$

*and t stored at a node* $\mathcal{S}_i \in \overline{\mathsf{Z}}$ *is semi-complete* if it contains all list elements $[\mathcal{S}_j,\mathbf{O},t_1,t_2]$ with $t_1 \leq t \leq t_2$ where $\mathcal{S}_j \in \mathsf{DetNeigh}_i^{\mathsf{Z}}$. □

**Lemma 27.** *Let* $\mathcal{S}_i$ *detect* $\mathbf{O}$ *at* $t$. *Without loss of generality, let* $\mathcal{S}_i \in \mathsf{Z}$. *If* Detections *stored at* $\mathcal{S}_i$ *is semi-complete regarding* $\mathbf{O}$ *and* $t$, *the computation of the detection scenario at* $\mathcal{S}_i$ *according to Table 5 is correct.*

*Proof.* Since $\mathcal{S}_i$ detects $\mathbf{O}$, one can deduct $isDetecting(\mathsf{Z}, t, \mathbf{O}) = \mathcal{T}$. Thus, only $isDetecting\left(\overline{\mathsf{Z}}, t, \mathbf{O}\right)$ remains to be computed by $\mathcal{S}_i$. Computing $isDetecting\left(\overline{\mathsf{Z}}, t, \mathbf{O}\right)$ only requires list elements originating from nodes in $\overline{\mathsf{Z}}$. ■

Lemma 27 implies that the detection scenario computation is still correct if Detections contains only list elements from a subset of certain detection neighbors. This reduces the number of messages, in particular because this subset is empty for most nodes.

**Definition 42 (Border Node):** A *border node* is

- a node $\mathcal{S}_i \in \mathsf{Z}$ with $\mathsf{DetNeigh}_i^{\overline{\mathsf{Z}}} \neq \varnothing$, or
- a node $\mathcal{S}_i \in \overline{\mathsf{Z}}$ with $\mathsf{DetNeigh}_i^{\mathsf{Z}} \neq \varnothing$. □

Figure 8 illustrates the concept of border nodes: Non-border nodes inside Z are represented by black-colored circles. Black-colored squares correspond to border nodes inside Z. Similarly, gray-colored squares and circles correspond to border and non-border nodes outside of Z, respectively. A significant share of the nodes in this scenario are non-border nodes. This is important, because non-border nodes compute detection scenarios without obtaining elements for Detections originating from any detection neighbor.

**Lemma 28.** *If a non-border node* $\mathcal{S}_i$ *detects* $\mathbf{O}$ *at* $t$ *and modifies* Detections *accordingly,* Detections *stored at* $\mathcal{S}_i$ *is semi-complete.*

*Proof.* Without loss of generality let $\mathcal{S}_i \in \mathsf{Z}$ and $\mathsf{DetNeigh}_i^{\overline{\mathsf{Z}}} = \varnothing$, i.e., $\mathcal{S}_i$ is not a border node. $\mathsf{DetNeigh}_i^{\overline{\mathsf{Z}}} = \varnothing$ implies that there does not exist a node $\mathcal{S}_j \in \overline{\mathsf{Z}}$ whose detection area overlaps with the detection area of $\mathcal{S}_i$. Thus, simultaneous detection of an object by $\mathcal{S}_i$ and some $\mathcal{S}_j \in \overline{\mathsf{Z}}$ is not possible by definition. Hence, detection of an object $\mathbf{O}$ by $\mathcal{S}_i$ implies $isDetecting\left(\overline{\mathsf{Z}}, t, \mathbf{O}\right) = \mathcal{F}$ and $isDetecting(\mathsf{Z}, t, \mathbf{O}) = \mathcal{T}$. ■

Depending on the structure of the development in question, the concept of border nodes allows for further reduction of communication, as follows:

**Lemma 29.** *Let* $\mathbb{P}(\mathbf{O},\mathsf{Z}) = P_1(\mathbf{O},\mathsf{Z}) \triangleright P_2(\mathbf{O},\mathsf{Z})$. *Detections and the resulting list elements stored in* Detections *originating from non-border nodes are not necessary to process* $\mathbb{P}(\mathbf{O},\mathsf{Z})$.

*Proof.* Without loss of generality, assume the non-border node $\mathcal{S}_i$ detects $\mathbf{O}$ at time $t_1$ and derives a detection scenario $\mathsf{DS}^*$ that yields $\mathrm{P}_1(\mathbf{O},\mathsf{Z}) = \mathcal{T}$

according to Table 3. If $\mathbf{O}$ fulfills $\mathbb{P}(\mathbf{O}, \mathsf{Z})$ at some time $t_2 > t_1$, there will be a border node $\mathcal{S}_j$ that detects $\mathbf{O}$ and computes $\mathsf{DS}^*$. Thus, $\mathcal{S}_j$ derives $\mathrm{P}_1(\mathbf{O}, \mathsf{Z}) = \mathcal{T}$ as well and if $\mathbf{O}$ fulfills $\mathbb{P}(\mathbf{O}, \mathsf{Z})$ this must be followed directly by $\mathrm{P}_2(\mathbf{O}, \mathsf{Z}) = \mathcal{T}$. If no such border node exists, $\mathbf{O}$ does not fulfill $\mathbb{P}(\mathbf{O}, \mathsf{Z})$ and therefore $\mathbf{O}$ is irrelevant regarding the users interest. Hence, the detection of a non-border node is irrelevant for developments like $\mathbb{P}(\mathbf{O}, \mathsf{Z})$. ■

Summing up, non-border nodes are inactive when developments like $\mathrm{Enter}(\mathbf{O}, \mathsf{Z})$ are processed.

### 8.4.1 Approximation of Detection Neighbors

There exist detection mechanisms where the detection area is indeterminable, i.e., nodes cannot determine their detection neighbors. We solve this by using a superset $\mathsf{ApproxDetNeigh}_i$ which contains at least all detection neighbors $\mathsf{DetNeigh}_i$, i.e., $\mathsf{DetNeigh}_i \subseteq \mathsf{ApproxDetNeigh}_i$. Results obtained while using $\mathsf{ApproxDetNeigh}_i$ instead of $\mathsf{DetNeigh}_i$ are still correct, because those nodes in $\mathsf{ApproxDetNeigh}_i$ that are not detection neighbors of $\mathcal{S}_i$ cannot detect an object at the same time as $\mathcal{S}_i$. Several approaches to derive such a superset are conceivable, and we outline two of them:

**Communication Neighbors**: If the communication range can be assumed to be much larger than the maximum detection range, a valid superset is $\mathsf{CN}_i$, i.e., $\mathsf{ApproxDetNeigh}_i = \mathsf{CN}_i$. This approach is applicable to most detection mechanisms used in SN, and we use it for our evaluation.

**Node Positions**: Another approach is applicable if the nodes know their position: The set $\mathsf{ApproxDetNeigh}_i$ contains all nodes with a distance of at most $2 \cdot \mathcal{D}_{max}$ to $\mathcal{S}_i$. The factor 2 ensures that the circles around $\mathsf{POS}_i$ with radius $\mathcal{D}_{max}$ do not overlap.

Next, we propose two strategies to obtain the list elements for Detections originating from detection neighbors.

### 8.4.2 Reactive Data Collection

Algorithm 8.2 outlines the *reactive strategy*, and its core idea is as follows: According to Table 3, for each predicate $\mathrm{P}(\mathbf{O}, \mathsf{Z})$ there is one detection scenario where $\mathrm{P}(\mathbf{O}, \mathsf{Z}) = \mathcal{T}$. Thus, each node can determine which predicates and thus the detection scenarios that are relevant to process the query. For instance, for $\mathrm{P}(\mathbf{O}, \mathsf{Z}) = \mathrm{SNEnter}(\mathbf{O}, \mathsf{Z})$ each node knows that only $\mathsf{DS}^E$ and $\mathsf{DS}^I$ are relevant. When an object $\mathbf{O}$ enters or leaves the detection area of $\mathcal{S}_i$ at time $t$, $\mathcal{S}_i$ checks if this possibly results in a predicate $\mathrm{P}(\mathbf{O}, \mathsf{Z})$ of the query being true. If so, $\mathcal{S}_i$ requests Detections-entries on $\mathbf{O}$ from a subset of its detection neighbors. We denote this subset as $\mathsf{DetNeigh}_i^{sub}$. After receiving and storing the entries requested from $\mathsf{DetNeigh}_i^{sub}$, $\mathcal{S}_i$ computes the detection

**Algorithm 8.2:** Reactive Strategy

```
When O enters or leaves the detection area of S_i do
    Modify Detections as described in Section 8.1;
    DetNeigh_i^sub ← Detection neighbors that must be queried according to Table 6;
    Request tuples on O from every node in DetNeigh_i^sub;
    Wait for response from every node in DetNeigh_i^sub;
    Determine detection scenario according to Table 5;
    Notify base station if O fulfills a predicate of the query according to Table 3;
end
```

| Reactive | | $\mathcal{S}_i \in \mathsf{Z}$ | $\mathcal{S}_i \in \overline{\mathsf{Z}}$ |
|---|---|---|---|
| $\mathsf{DS}^I$ | Entry | $\mathsf{DetNeigh}_i^{\overline{\mathsf{Z}}}$ | $\varnothing$ |
| | Exit | $\varnothing$ | $\mathsf{DetNeigh}_i$ |
| $\mathsf{DS}^B$ | Entry | $\mathsf{DetNeigh}_i^{\overline{\mathsf{Z}}}$ | $\mathsf{DetNeigh}_i^{\mathsf{Z}}$ |
| | Exit | $\varnothing$ | $\varnothing$ |
| $\mathsf{DS}^E$ | Entry | $\varnothing$ | $\mathsf{DetNeigh}_i^{\mathsf{Z}}$ |
| | Exit | $\mathsf{DetNeigh}_i$ | $\varnothing$ |

**Table 6** Detection-neighbor partitions for the reactive strategy

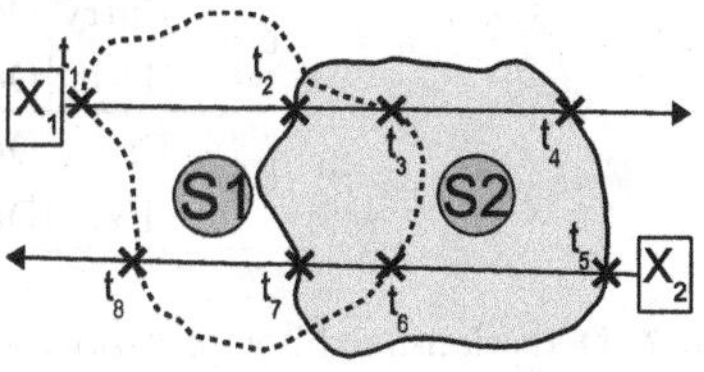


**Fig. 20** Detection Events ($\mathcal{S}_1 \in \overline{\mathsf{Z}}$, $\mathcal{S}_2 \in \mathsf{Z}$)

scenario as described above. If the detection scenario computed indicates that a predicate in $\mathbb{P}(\mathbf{O},\mathsf{Z})$ is true, the base station is notified. The core question is: "When $\mathcal{S}_i$ detects $\mathbf{O}$, which detection neighbors could have tuples that are relevant to compute the detection scenario, i.e., which nodes must be in $\mathsf{DetNeigh}_i^{sub}$?"

We can derive $\mathsf{DetNeigh}_i^{sub}$ based on (1) the detection scenario to compute, (2) whether $\mathbf{O}$ entered or left the detection area of $\mathcal{S}_i$ and (3) whether $\mathcal{S}_i$ is in $\mathsf{Z}$ or $\overline{\mathsf{Z}}$. Table 6 shows $\mathsf{DetNeigh}_i^{sub}$ for any combination of these three parameters. We use Figure 20 to explain each entry: The first row is related to $\mathsf{DS}^I$, i.e., $\mathbb{P}(\mathbf{O},\mathsf{Z})$ contains Inside $(\mathbf{O},\mathsf{Z})$. $\mathsf{DS}^I$ can only occur (1) if $\mathbf{O}$ enters the detection area of a node in $\mathsf{Z}$, or (2) if $\mathbf{O}$ leaves the detection area of a node in $\overline{\mathsf{Z}}$. Case (1) is illustrated at $t_2$ and $t_5$ in Figure 9. To determine if $\mathsf{DS}^I$ has occurred, $\mathcal{S}_2$ only has to communicate with detection neighbors outside of $\mathsf{Z}$, i.e., with $\mathcal{S}_1$. This is reflected by $\mathsf{DetNeigh}_i^{\overline{\mathsf{Z}}}$ in Table 6. Case (2) occurs at $t_3$ and $t_8$ in Figure 20. The node whose detection area the object has left must send a request to all detection neighbors, i.e., $\mathsf{DetNeigh}_i$. This is because there must be at least one detection neighbor in $\mathsf{Z}$ and no detection neighbor outside of $\mathsf{Z}$ that still detects the object. The reasoning for $\mathsf{DS}^E$ is analogous. $\mathsf{DS}^B$ only occurs if nodes inside and outside of $\mathsf{Z}$ are in $\mathsf{DetSet}_t^{\mathbf{O}}$. Hence, any $\mathcal{S}_i \in \mathsf{Z}$ requests only tuples from $\mathsf{DetNeigh}_i^{\overline{\mathsf{Z}}}$ and vice versa.

### 8.4.3 Proactive Data Collection

The core idea of the proactive strategy is that a nodes whose detection area was entered or left by an object send this information to some of their detection neighbors (cf. Algorithm 8.3). This allows each receiver of the update to check if a predicate of the query was true. Algorithm 8.3 illustrates the strategy.

| **Proactive** | | $\mathcal{S}_i \in \mathsf{Z}$ | $\mathcal{S}_i \in \overline{\mathsf{Z}}$ |
|---|---|---|---|
| $\mathsf{DS}^I$ | Entry | $\varnothing$ | $\mathsf{DetNeigh}_i^{\mathsf{Z}}$ |
| | Exit | $\varnothing$ | $\mathsf{DetNeigh}_i^{\mathsf{Z}}$ |
| $\mathsf{DS}^B$ | Entry | $\mathsf{DetNeigh}_i^{\overline{\mathsf{Z}}}$ | $\mathsf{DetNeigh}_i^{\mathsf{Z}}$ |
| | Exit | $\mathsf{DetNeigh}_i^{\overline{\mathsf{Z}}}$ | $\mathsf{DetNeigh}_i^{\mathsf{Z}}$ |
| $\mathsf{DS}^E$ | Entry | $\mathsf{DetNeigh}_i^{\overline{\mathsf{Z}}}$ | $\varnothing$ |
| | Exit | $\mathsf{DetNeigh}_i^{\overline{\mathsf{Z}}}$ | $\varnothing$ |

**Table 7** Detection-neighbor partitions for the proactive strategy

The computation of $\mathsf{DetNeigh}_i^{sub}$ is the most important step. As with the reactive strategy, it depends on the three aforementioned parameters which detection neighbors must be in $\mathsf{DetNeigh}_i^{sub}$. For any combination of these parameters, there is an entry in Table 7. We briefly explain each entry: Recall that $\mathsf{DS}^I$ can either occur (1) when an object enters the detection area of a node in $\mathsf{Z}$ or (2) when the detection area of a node in $\overline{\mathsf{Z}}$ is left. Using Figure 20 again, Case (1) occurs at $t_2$ and $t_5$. To compute the detection scenario correctly at $t_2$, $\mathcal{S}_2$ must know that $\mathcal{S}_1 \in \overline{\mathsf{Z}}$ currently detects $\mathbf{X}_1$. Case (2) occurs when $\mathbf{X}_1$ leaves the detection area of $\mathcal{S}_1$ at $t_3$ and $t_8$. Thus, if the query requires $\mathsf{DS}^I$, nodes in $\mathsf{Z}$ must receive updates from their detection neighbors outside of $\mathsf{Z}$, i.e., $\mathsf{DetNeigh}_i^{\overline{\mathsf{Z}}}$. The entries for $\mathsf{DS}^E$ are explained

**Algorithm 8.3:** Proactive Strategy

1 **When O** *enters/leaves detection area of* $\mathcal{S}_i$ **do**
2 Modify $\mathsf{Detections}$ as described in Section 8.1;
3 $\mathsf{DetNeigh}_i^{sub} \leftarrow$ detection neighbors whose information must be updated according to Table 7;
4 Send updated list entries to every node in $\mathsf{DetNeigh}_i^{sub}$;
5 **Goto** Line 9;
6 **end**
7 **When** $\mathcal{S}_i$ *receives updated tuples about* **O do**
8 Insert updated tuples into $\mathsf{Detections}$;
9 Determine detection scenario according to Table 5;
10 Notify base station if **O** fulfills a predicate of the query;
11 **end**

analogously: Nodes outside of Z must be informed about object detections of their detection neighbors inside Z, i.e., $\mathsf{DetNeigh}_i^{\mathsf{Z}}$. $\mathsf{DS}^B$ requires simultaneous detection by nodes in Z as well as $\overline{\mathsf{Z}}$. Thus, every $\mathcal{S}_i \in \mathsf{Z}$ must be informed about detections of detection neighbors in $\overline{\mathsf{Z}}$ and vice versa.

## *8.5 Impact of Node Failures*

Node failures can have two consequences: (1) An object **O** that would have been detected is not detected. (2) Nodes detect **O**, but the detection-scenario computation is possibly incorrect because it is based on an incomplete list Detections. Section 7 has shown how users can express queries if they are interested in objects that are temporarily unobserved. Thus, we focus on (2), i.e., we notify the user if query results returned could be incorrect due to node failures. We discuss the detection of failures first and continue with failure handling.

### 8.5.1 Failure Detection

The reactive strategy implicitly supports failure detection because a node $\mathcal{S}_i$ expects responses from a set of detection neighbors $\mathsf{DetNeigh}_i^{sub}$. $\mathcal{S}_i$ can derive that detection neighbors which have not sent such a response after a timeout have failed.

Failure detection requires additional measures with the proactive strategy, because failed detection neighbors cannot be identified based on missing responses. One such measure is sending beacon messages periodically and assuming node failures if beacons are missing. We include this overhead in our evaluation. This problem also occurs with the centralized strategy, i.e., additional messages are required to detect node failures.

### 8.5.2 Failure Handling

Our goal is to notify the user if a failure could have an impact on the query result and mark the corresponding result accordingly. We refer to the node whose failure has been detected as $\mathcal{S}_{fail}$. Any detection scenario $\mathsf{DS}_{err}$ computed by a node $\mathcal{S}_i$ with $\mathcal{S}_{fail} \in \mathsf{DetNeigh}_i$ may be incorrect because the list Detections was incomplete.

**Lemma 30.** *If* $\mathsf{DS}_{err} = \mathsf{DS}^B$, *the failure of* $\mathcal{S}_{fail}$ *did not affect the computation of the detection scenario.*

*Proof.* According to Table 3, $\mathsf{DS}^B$ occurs if there exists at least one node in Z and one node outside of Z that detect the object. This is independent from the potential detection of $\mathcal{S}_{fail}$ and thus the detection scenario computation is not affected by the failure of $\mathcal{S}_{fail}$. ■

**Lemma 31.** *If* $\mathsf{DS}_{err} = \mathsf{DS}^I$ *and* $\mathcal{S}_{fail} \in \mathsf{Z}$ *or* $\mathsf{DS}_{err} = \mathsf{DS}^E$ *and* $\mathcal{S}_{fail} \in \overline{\mathsf{Z}}$, *the failure of* $\mathcal{S}_{fail}$ *did not affect the computation of the detection scenario.*

*Proof.* We prove this only for the case of $\mathsf{DS}_{err} = \mathsf{DS}^I$. The reasoning for $\mathsf{DS}_{err} = \mathsf{DS}^E$ is analogous. $\mathsf{DS}_{err} = \mathsf{DS}^I$ implies that there exists a node $\mathcal{S}_j \in \mathsf{Z}$ that currently detects $\mathbf{O}$. Since $\mathcal{S}_{fail} \in \mathsf{Z}$, an additional list entry originating from $\mathcal{S}_{fail}$ would not change the result of the detection scenario computation. Hence, the failure of $\mathcal{S}_{fail}$ cannot affect the result. ■

We infer that the base station must be notified only if one of the following two cases occurs:

- $\mathsf{DS}_{err} = \mathsf{DS}^I$, and $\mathcal{S}_{fail} \in \overline{\mathsf{Z}}$
- $\mathsf{DS}_{err} = \mathsf{DS}^E$, and $\mathcal{S}_{fail} \in \mathsf{Z}$

If one of these cases occurs, the notification to the base station contains $\mathsf{DS}_{err}$ and $\mathcal{S}_{fail}$.

# 9 Evaluation

Our evaluation investigates the communication required for spatio-temporal query processing. Communication is the most important factor regarding the lifetime of battery-powered nodes. We address the following hypotheses:

**H1** The centralized strategy does not scale as well as the distributed strategies regarding network size and node density.

**H2** For Inside $(\mathbf{O}, \mathsf{Z})$ and Disjoint $(\mathbf{O}, \mathsf{Z})$, the proactive strategy is most-energy efficient.

**H3** The reactive strategy is the most energy-efficient for Meet $(\mathbf{O}, \mathsf{Z})$.

**H4** Distributed strategies reduce communication required for processing spatio-temporal developments.

We use the number of messages as a proxy for communication. This is justified in Appendix A. It analyses the energy consumption of sensor nodes and energy measurements with Sun SPOT [50] and Mica [55] nodes. In the following, we present results from experiments using simulations as well as Sun SPOT deployments.

## *9.1 Simulation Configuration*

We have used the KSN Sun SPOT simulator [6] to obtain our results. This is because it allows the usage of the same software for simulations as well as Sun SPOT deployments. Each simulation run consists of the following steps:

1. Randomly deploy 100-300 nodes over an area of constant size. Using an area of constant size ensures different node densities for different network sizes, i.e., varying numbers of detection and communication neighbors.
2. Define a zone containing between 2 and 30 nodes.
3. Generate 50 different object paths using a random walk model with starting points randomly chosen.
4. For each object path, evaluate each detection scenario using each strategy.

The results presented here are based on more than 100.000 simulation runs.

Recall that determining the detection area is not possible for certain detection mechanisms. Our simulations take this into account by using the set of communication neighbors as a proxy for the set of detection neighbors: Every node sends a beacon periodically, and every receiver of such a beacon adds the node to the list of detection neighbors. We report the additional communication related to beacons separately. Note that in case of proactive data collection, the beacon messages can be used for failure detection as well.

## *9.2 Simulation Results*

Figure 21 shows that **H1** holds: The graph plots the average number of messages per simulation run to compute Inside $(\mathbf{O}, \mathsf{Z})$. We omit similar graphs for the other predicates. The number of messages required by the centralized strategy increases linearly with network size. This is expected, because objects are detected by more nodes. This results in many messages which must be forwarded to the base station. Despite the additional communication to approximate detection neighbors, the increasing network size (and node density) affects both distributed strategies only marginally. We leave more sophisticated measures to determine detection neighbors more efficiently for future work.

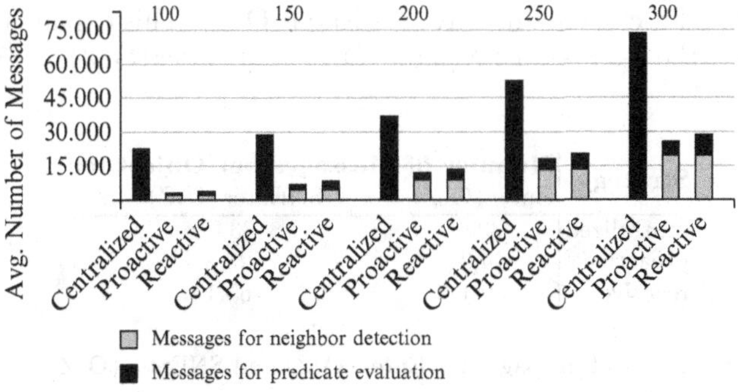


**Fig. 21** Scalability of data-collection strategies for Inside $(\mathbf{O}, \mathsf{Z})$ for SN with 100-300 nodes

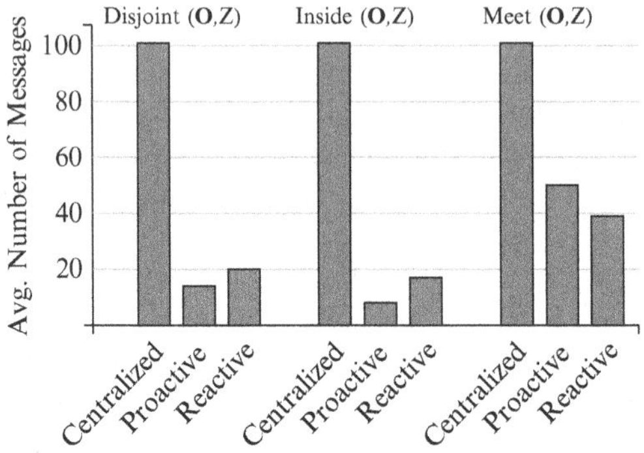


**Fig. 22** Comparison of communication costs for the evaluation of different detection scenarios

To investigate **H2** and **H3**, we compare the average number of messages required to process a given predicate. The result in Figure 22 indicates that distributed strategies reduce communication by $45\% - 85\%$, compared to the centralized strategy. As expected, the proactive strategy is advantageous for Inside $(\mathbf{O}, \mathsf{Z})$ and Disjoint $(\mathbf{O}, \mathsf{Z})$. This is because $\mathsf{S}^*$ is smaller for the proactive strategy compared to the reactive one when objects leave a detection area of a node $\mathcal{S}_i \in \overline{\mathsf{Z}}$ (cf. Tables 6 and 7). These roles are reversed for $\mathtt{DS}^B$, because the proactive strategy is triggered more often than the reactive one. These results confirm **H2** and **H3**.

Our simulation results support **H4** as well: Table 8 shows the average number of messages to determine that $\mathbf{O}$ conforms to Enter $(\mathbf{O}, \mathsf{Z})$ or SNEnter $(\mathbf{O}, \mathsf{Z})$ (cf. equations (29) and (33)), respectively. Compared to the centralized strategy, the distributed strategies reduce communication by 51% to 89%. This is because only few nodes send messages to the base station via multiple hops. The proactive strategy is most efficient, because SNEnter $(\mathbf{O}, \mathsf{Z})$ does not contain Meet $(\mathbf{O}, \mathsf{Z})$. According to Lemma 29, all non-border nodes can stay inactive for Enter $(\mathbf{O}, \mathsf{Z})$. This explains the difference between the results for SNEnter $(\mathbf{O}, \mathsf{Z})$ and Enter $(\mathbf{O}, \mathsf{Z})$.

| **Strategy** | **Number of Messages per Object for** | |
|---|---|---|
| | Enter $(\mathbf{O}, \mathsf{Z})$ | SNEnter $(\mathbf{O}, \mathsf{Z})$ |
| centralized | 334 | 334 |
| proactive | 44,3 | 123,8 |
| reactive | 39,1 | 163,1 |

**Table 8** Avg. number of messages for Enter $(\mathbf{O}, \mathsf{Z})$ and SNEnter $(\mathbf{O}, \mathsf{Z})$

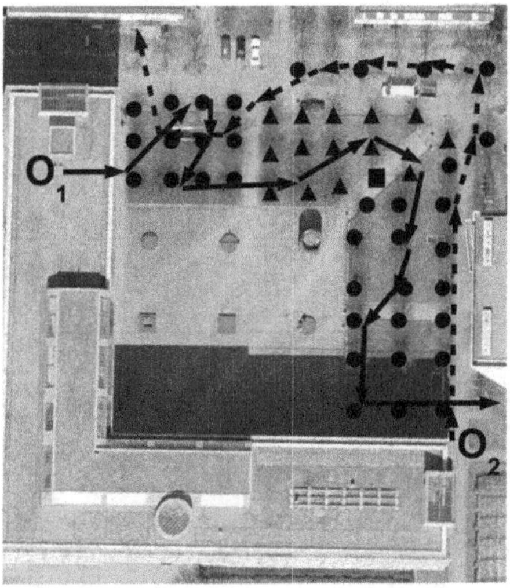


**Fig. 23** Node distribution and object paths for the Sun SPOT case study conducted at the Karlsruhe Institute of Technology

## *9.3 Sun SPOT Case Study*

Simulations abstract from certain real-world phenomena which may impact performance, e.g., interferences or collisions. We have conducted a case study using several indoor and outdoor deployments to confirm our simulation results in a real SN. We now present the most important results.

The node distribution of an outdoor deployment is shown in Figure 23: We have mounted 50 Sun SPOT sensor nodes on trees over an area of more than $2500m^2$ and positioned the base station (black square) in the middle of the SN. Nodes inside and outside of Z are represented by triangles and circles respectively. Two remote controlled cars $\mathbf{O}_1$ (solid line) and $\mathbf{O}_2$ (dashed line) moved through the SN. We have processed the developments Enter $(\mathbf{O}_1, \mathrm{Z})$ and Touch $(\mathbf{O}_2, \mathrm{Z})$ with all strategies and counted the messages required.

The centralized strategy always requires significantly more communication than the other two strategies. This is surprising, because the experiment favors the centralized strategy: Experiments with the base station at the border of the SN result in even larger differences compared to the distributed strategies. Objects that are detected by more nodes or that move in patterns intensify this effect. Considering Tables 6 and 7, the results of the distributed strategies are expected: As shown above, the proactive strategy requires less communication for Disjoint $(\mathbf{O}_1, \mathrm{Z})$ and Inside $(\mathbf{O}_1, \mathrm{Z})$ than the reactive one. For Meet $(\mathbf{O}_1, \mathrm{Z})$, these roles are reversed. The proactive strategy is better for Enter $(\mathbf{O}_1, \mathrm{Z})$ than the reactive strategy because Meet $(\mathbf{O}_1, \mathrm{Z})$ only occurs a few times. This is different for $\mathbf{O}_2$: Meet $(\mathbf{O}_2, \mathrm{Z})$ is true for a long time due to the trajectory of $\mathbf{O}_2$ which results in many messages with the proactive strategy. This confirms our simulation results. We conclude that the evaluation supports all hypotheses.

| Strategy | Number of Messages | |
|---|---|---|
| | Enter $(\mathbf{O}_1, Z)$ | Touch $(\mathbf{O}_2, Z)$ |
| centralized | 264 | 302 |
| proactive | 159 | 217 |
| reactive | 184 | 178 |

**Table 9** Results for Enter $(\mathbf{O}_1, Z)$ and Touch $(\mathbf{O}_2, Z)$

## 10 Conclusions

There exist many SN applications that track moving objects. While research has shown that accessing SN declaratively is advantageous, only relational queries have been addressed so far. Relational operators are insufficient to express the spatio-temporal semantics required by applications that track objects. This paper provides the foundations for spatio-temporal queries in SN. By developing an abstract detection model and introducing the concept of detection scenarios, we have formalized the information obtained by object detection mechanisms in SN. Furthermore, we systematically investigated different types of spatio-temporal queries in SN: Based on detection scenarios, we have translated object detections into results for every type of query. There are cases where the inaccuracy of object detection is in the way of a definite answer to a query, i.e., the query result is approximate. As we have proven, the results derived by our techniques are optimal in these cases. As a last step, we have proposed concepts for energy-efficient in-network processing of spatio-temporal queries in SN. We have evaluated our measures using simulations as well as real deployments of sensor nodes. The results how that our in-network processing reduces communication by 45%-89% compared to collecting all information on object detections at the base station.

## References

[1] Abadi, D.J., et al.: REED: Robust, efficient Filtering and Event Detection in Sensor Networks. In: VLDB (2005)

[2] Advantaca, Inc.: TWR-ISM-002-I Radar: Hardware User's Manual (2002)

[3] Ahmed, N., et al.: The holes problem in wireless sensor networks: a survey. SIGMOBILE Mob. Comput. Commun. Rev. (2005)
[4] de Almeida, V.T., Güting, R.H.: Supporting uncertainty in moving objects in network databases. In: GIS '05 (2005)
[5] Arora, A., et al.: A line in the sand: A wireless sensor network for target detection, classification, and tracking. Computer Networks (2004)
[6] Bestehorn, M., et al.: The Karlsruhe Sensor Networking Project (KSN) (2007). URL `http://www.ipd.kit.edu/KSN`
[7] Bestehorn, M., et al.: Deriving Spatio-temporal Query Results in Sensor Networks. In: SSDBM (2010)
[8] Bestehorn, M., et al.: Energy-efficient processing of spatio-temporal queries in wireless sensor networks. In: ACM SIGSPATIAL GIS (2010)
[9] Bonnet, P., et al.: Querying the Physical World. Personal Communications, IEEE (2000)
[10] Bonnet, P., et al.: Towards sensor database systems. In: MDM '01 (2001)
[11] Braunling, R., et al.: Acoustic Target Detection, Tracking, Classification, and Location in a Multiple-Target Environment. In: SPIE (1997)
[12] Buettner, M., et al.: X-mac: a short preamble mac protocol for duty-cycled wireless sensor networks. In: SenSys '06 (2006)
[13] Cao, H., et al.: Spatio-temporal data reduction with deterministic error bounds. VLDB J. **15** (2006)
[14] Cerpa, A., et al.: Habitat monitoring: Application driver for wireless communications technology. SIGCOMM CCR (2001)
[15] Chu, D., et al.: Approximate data collection in sensor networks using probabilistic models. In: ICDE '06 (2006)
[16] Ding, J., et al.: Signal Processing of Sensor Node Data for Vehicle Detection. In: IEEE ITSC (2004)
[17] Dutta, P.K., et al.: Towards radar-enabled sensor networks. In: IPSN '06 (2006)
[18] Egenhofer, M.J., Franzosa, R.D.: Point set topological relations. IJGIS (1991)
[19] Erwig, M., Schneider, M.: Spatio-temporal predicates. IEEE TKDE (2002)
[20] Fonseca, R., et al.: The collection tree protocol (ctp) (2007). URL `http://www.tinyos.net/tinyos-2.x/doc/html/tep123.html`
[21] Gaal, S.: Point set topology. Academic Press (1964)
[22] Gamage, C., et al.: Security for the mythical air-dropped sensor network. In: ISCC (2006)
[23] Gehrke, J., Madden, S.: Query processing in sensor networks. Pervasive Computing, IEEE (2004)
[24] Grilo, A., et al.: A wireless sensor network architecture for homeland security application. In: ADHOC-NOW (2009)
[25] Güting, R.H., et al.: A Foundation for Representing and Querying Moving Objects. ACM TODS (2000)

[26] Güting, R.H., et al.: Modeling and querying moving objects in networks. VLDB J. (2006)
[27] He, T., et al.: Energy-efficient surveillance system using wireless sensor networks. In: MobiSys '04 (2004)
[28] He, T., et al.: Vigilnet: An integrated sensor network system for energy-efficient surveillance. ACM Trans. Sen. Netw. **2** (2006)
[29] Hergenröder, A., Wilke, J., Meier, D.: Distributed Energy Measurements in WSN Testbeds with a Sensor Node Management Device (SNMD) (2010)
[30] Hill, J., et al.: System architecture directions for networked sensors. SIGPLAN Not. **35**(11) (2000)
[31] Klues, K., et al.: A component-based architecture for power-efficient media access control in wireless sensor networks. In: SenSys '07 (2007)
[32] Knuth, D.E., et al.: Fast Pattern Matching in Strings. SIAM Journal on Computing (1977)
[33] Koenig, W., et al.: Detectability, Philopatry, and the Distribution of Dispersal Distances in Vertebrates. Trends in Ecology & Evolution (1996)
[34] Kung, H., Vlah, D.: Efficient location tracking using sensor networks. IEEE WCNC (2003)
[35] Langendorfer, P., et al.: A Wireless Sensor Network Reliable Architecture for Intrusion Detection. In: NGI (2008)
[36] Li, D., et al.: Detection, Classification, and Tracking of Targets. Signal Processing Magazine, IEEE (2002)
[37] Liu, N.H., et al.: Long-term animal observation by wireless sensor networks with sound recognition. In: WASA '09 (2009)
[38] Liu, T., et al.: Implementing Software on Resource-Constrained Mobile Sensors: Experiences with Impala and ZebraNet. In: MobiSys '04 (2004)
[39] Madden, S., et al.: Tag: a tiny aggregation service for ad-hoc sensor networks. SIGOPS OSDI (2002)
[40] Madden, S., et al.: The design of an acquisitional query processor for sensor networks. In: SIGMOD '03 (2003)
[41] Madden, S., et al.: TinyDB: An Acquisitional Query Processing System for Sensor Networks. ACM TODS (2005)
[42] Madden, S.R.: The design and evaluation of a query processing architecture for sensor networks. Ph.D. thesis, University of California at Berkeley, Berkeley, CA, USA (2003). Chair-Franklin, Michael J.
[43] Mainwaring, A., et al.: Wireless sensor networks for habitat monitoring. In: WSNA (2002)
[44] Metsaranta, J.M.: Assessing Factors Influencing the Space Use of a Woodland Caribou Rangifer Tarandus Caribou Population using an Individual-Based Model. Wildlife Biology (2008)
[45] Perkins, C.E., et al.: Internet Connectivity for Ad Hoc Mobile Networks (2002)
[46] Polastre, J., et al.: Versatile low power media access for wireless sensor networks. In: SenSys '04 (2004)

[47] Rettie, J.W., Messier, F.: Hierarchical Habitat Selection by Woodland Caribou: Its Relationship to Limiting Factors. Ecography (2000)
[48] Shrivastava, N., et al.: Target tracking with binary proximity sensors: fundamental limits, minimal descriptions, and algorithms. In: SenSys '06 (2006)
[49] Succi, G.P., et al.: Acoustic target tracking and target identification: recent results. Unattended Ground Sensor Technologies and Applications (SPIE) (1999)
[50] SUN Microsystems Inc.: Small Programmable Object Technology (SPOT) (2009)
[51] Tilove, R.B.: Set Membership Classification: A Unified Approach to Geometric Intersection Problems. IEEE TC (1980)
[52] Trajcevski, G., et al.: The geometry of uncertainty in moving objects databases. In: EDBT (2002)
[53] Trajcevski, G., et al.: Managing uncertainty in moving objects databases. ACM TODS (2004)
[54] Wolfson, O., et al.: Moving objects databases: Issues and solutions. SSDBM (1998)
[55] XBow Technology Inc.: Wireless sensor networks (2009)
[56] Yao, Y., Gehrke, J.: The Cougar Approach to In-Network Query Processing in Sensor Networks. SIGMOD Rec. (2002)
[57] Zhang, W., Cao, G.: Optimizing tree reconfiguration for mobile target tracking in sensor networks. INFOCOM 2004 (2004)

# Energy-Consumption in Sensor Networks

Prior to the energy efficient implementation of the query dissemination and spatio-temporal query processor, we conducted a set of experiments to determine the important aspects of energy efficiency in sensor networks. The setup and the results of these experiments are provided in this section and justify our approach to evaluate our measures by counting the number of messages sent and received. More precisely, the following hypotheses are proven experimentally:

**H.1** Exchanging information via wireless communication reduces the time until batteries are depleted significantly.

**H.2** Energy consumption for sending a message is marginally higher than receiving a message.

**H.3** The number of bytes contained in a single message has a minor impact on energy consumption, particularly if energy-efficient MAC protocols are used.

## A.1 Experimental Setup

To measure the energy consumption of a sensor node we used the ***S**ensor **N**ode **M**anagement **D**evice (SNMD)* [29]. The SNMD was attached between the battery and energy consuming components of the sensor nodes, e.g., CPU, memory, wireless communication chip and sensing board. Figure 24 illustrates a simplified circuit diagram of the setup and Figure 25 shows the energy measurement device with a Mica mote attached to it[1]. We measured the voltage drawn by the node at a high temporal resolution of up to 20 kHz and computed the energy consumption based on this.

[1] The experiments were conducted at a time when the SNMD was still in development and had a less compact appearance, but the basic features have not been changed since then.


M. Bestehorn, *Querying Moving Objects Detected by Sensor Networks*, SpringerBriefs in Computer Science, DOI 10.1007/978-1-4614-4927-0,

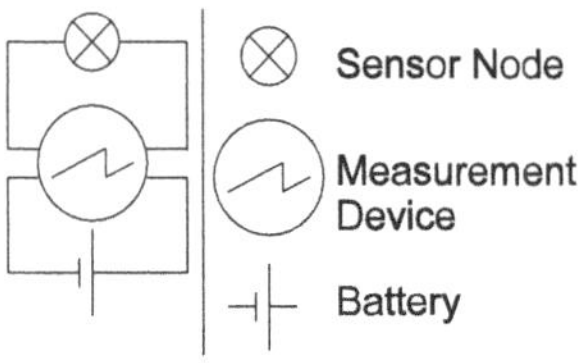


Fig. 24 Circuit diagram for energy measurements

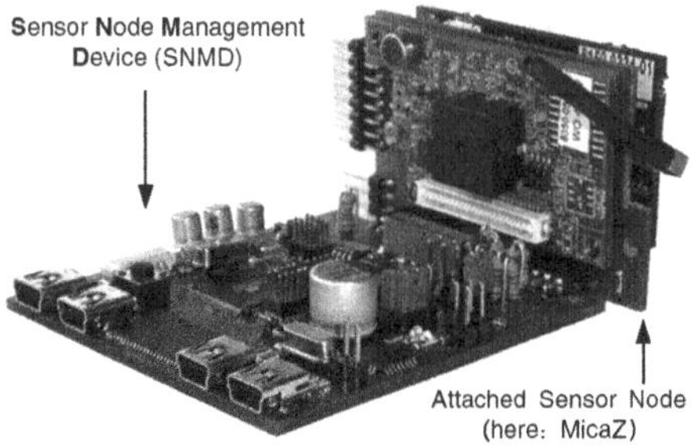


Fig. 25 Sensor Node Management Device [29] and attached Mica Mote

## A.2 Results and Analysis

For each measurement, we ran different applications with different properties regarding energy consumption on the nodes. We describe these applications first and the analyze the results of the energy measurement in the context of our hypotheses.

### *A.2.1 Impact of Communication on node lifetime*

For this experiment, we fully charged the batteries of three Sun SPOTs according to the specification of the battery. Afterwards, we assigned one of the following applications to one of the SPOTs:

**High:** This application prevented the usage of any power conservation features of the SPOTs. To increase power consumption, the radio and CPU were in use permanently.

**Medium:** This application raised an event on the SPOT every five minutes. After the event was raised, the SPOT sends data and then uses the *shallow sleep mode* to conserve energy while waiting for the next event. Note that the radio is not switched off during shallow sleep.

**Low:** A SPOT running this application is put into shallow sleep mode at all times.

While *shallow sleep mode* reduces energy consumption considerably, the overall power consumption is still orders of magnitude higher than in *deep sleep mode.* Table 10 compares both power saving modes.

SPOTs that run mainly in deep sleep mode can run for up to 900 days.

Figure 26 shows the measured voltage over time (in hours). The experiment ends when the battery reaches a critical voltage at $\approx 3.3V$. If this occurs, the battery hardware shuts the SPOT down. The application "high" depleted the battery an hour while each of the other applications ran 15 hours or more.

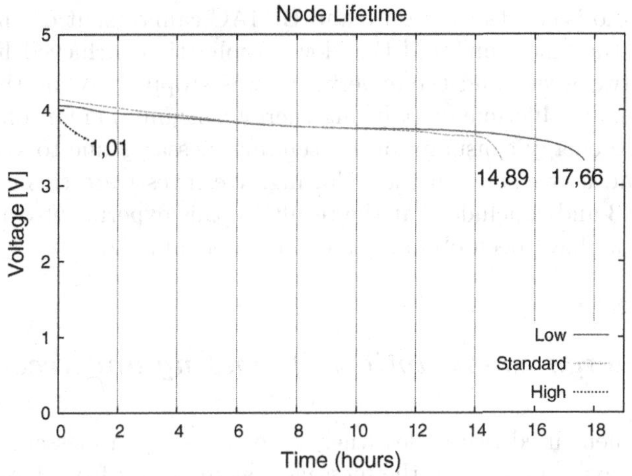


**Fig. 26** Node lifetime measurement result

| | Shallow Sleep | Deep Sleep |
|---|---|---|
| CPU | On with CPU clock stopped | Off |
| Master System Clock | On | Off |
| Low-level firmware | On | On[a] |
| RAM | On, but inactive | Main power off, RAM content preserved with low power standby |
| Flash memory | On, but inactive | Off |
| CC2240 radio chip | On | Off |
| AT91 peripheral clocks | On, if in use, otherwise off | Off |
| External/sensor board | On | Off |
| **Power consumption** | ≈ 24 mA | ≈ 32 $\mu A$ |

**Table 10** Energy saving modes for Sun SPOTs

[a] This is required to wake up the SPOT, e.g., at a given time.

Thus, constant usage of the CPU and radio drastically reduce the lifetime of the node. Comparing the "medium" and the "low" application shows that the additional use of the radio compared to using the sleep mode continuously reduces node lifetime considerably.

Regarding the absolute values in Figure 26 it must be noted, that in shallow sleep certain parts of the hardware are still switched on and consume energy as shown in Table 10. Most importantly, the radio is still switched on. Switching the radio off would require energy-aware MAC protocols, e.g., B-MAC [46], that ensure communication between nodes while switching off the radio for certain periods of time. Initial tests of the three applications with B-MAC resulted in the following: The "high" application was unaffected, since the radio is constantly in use and B-MAC cannot shut down the radio chip. Both, the "medium" and the "low" applications achieved lifetimes of more than two weeks and the experiment was stopped. While these protocols increase node lifetime by reducing energy consumption for idle listening, they increase energy consumption for sending messages due to synchronization overhead as we show in the following. We investigate this overhead in Section A.2.3 and conclude that the results of this experiment confirm **H.1**. Similar results have been obtained for Mica motes in [42].

### *A.2.2 Energy consumption of sending and receiving*

This experiment used two nodes where a sender sends a message of varying size to a receiver. The size of the message was increased from 1 packet to 10 packets. The experiment was conducted as follows:

1. Sender and receiver are started and switch off their radio.
2. The energy measurement using the SNMD is started and both SPOTs switch their radio back on.
3. After the radio is ready at the sender, the sender tries to deliver the first packet to the receiver.
4. After receiving an acknowledgement for the first packet, the next packet is sent. This procedure continues until all packets are sent.
5. After the last packet has been acknowledged by the receiver, the energy measurement is stopped.

Switching off the radio before the start of the experiment simulates the fact that before nodes can send messages the radio must be switched on. Keeping the radio on at all times is not a viable option as the experiment above has shown. Therefore the energy consumption for switching the radio on before sending a message must be taken into account to measure the energy consumed for sending a message. On both nodes, the usage of sleep modes or any other power-saving mechanism was prevented, i.e., all components of the nodes were on at any time. Both nodes used the default Sun SPOT MAC protocol, i.e., no energy-aware MAC protocol was used.

Figure 27 shows the result of the experiment. The difference regarding energy consumption between the sender and the receiver is marginal even for 10 packets. This confirms **H.2**. In addition to the result also shows that the size of the message has a minor impact on energy consumption even if there is no energy-aware MAC protocol. Sending a message consisting of a single packet consumed 9.14 mAs. Doubling the size to two packets leads to an energy consumption of 9.94 mAs, i.e., an increase of 8.8%. Increasing the message size by an order of magnitude only doubles the energy consumption. Thus, even without energy-aware MAC protocols **H.3** is confirmed. With energy-aware MAC protocols this relative increase becomes even smaller since these protocols induce a large constant overhead for sending and receiving. We investigate energy consumption of these protocols in the following.

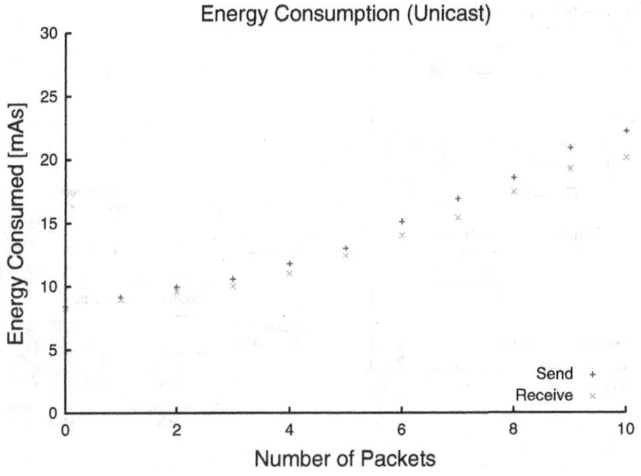


**Fig. 27** Energy Consumption for communication

## *A.2.3 Impact of energy-aware MAC protocols*

So far, all experiments used a default implementation of a 802.15.4 compatible MAC protocol which was not energy-aware, i.e., the radio chip was on at all times. This section investigates the impact of energy-aware MAC protocols such as B-MAC [46] or X-MAC [12]. Contrary to the previous experiments, we use Mica motes for this to ease presentation. This is because energy readings are difficult to interpret on Sun SPOTs because parallel processes such as garbage collection distort the readings. Experiments with SUN SPOTs had similar results which is expected since both node types use the CC2420 radio chip. Again, we used the Sensor Node Management Device (SNMD) [29] to obtain energy measurements.

We used two Mica Motes where the access to the wireless medium was controlled by a B-MAC implementation [31] provided with TinyOS [30]. One of the nodes broadcasts (sender) a single packet and the other mote (receiver) just receives the message broadcasted. Figure 28 shows the energy readings of both nodes and Figure 29 illustrates the schema of B-MAC.We explain the important points in time (marked with $[T_1], [T_2], [T_3], [T_4]$) for sender and receiver in the following.

The time interval $t$ at which each node checks the medium for incoming messages is 1 second. The point in time where sender and receiver switch on the radio and check if there are incoming messages are marked with $[T_1]$. For the first two intervals, both nodes switch off the radio immediately and save energy. After 2.3 seconds (at $[T_2]$), the sender application starts the sending process by switching the radio on, listening to the medium. Since there is no other node currently sending a message, the sender starts with the preamble.

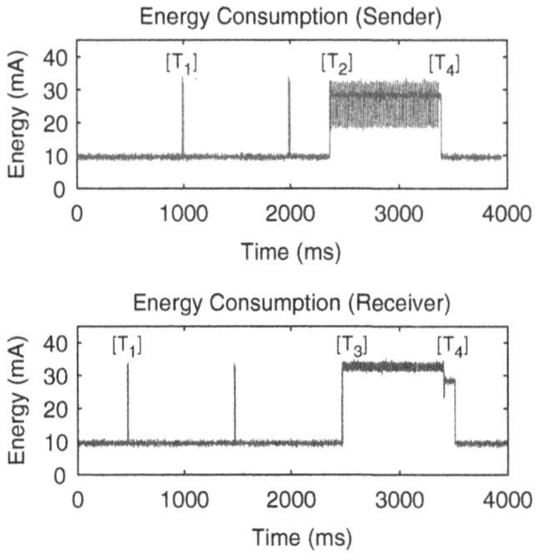


**Fig. 28** Energy consumption with B-MAC [46]

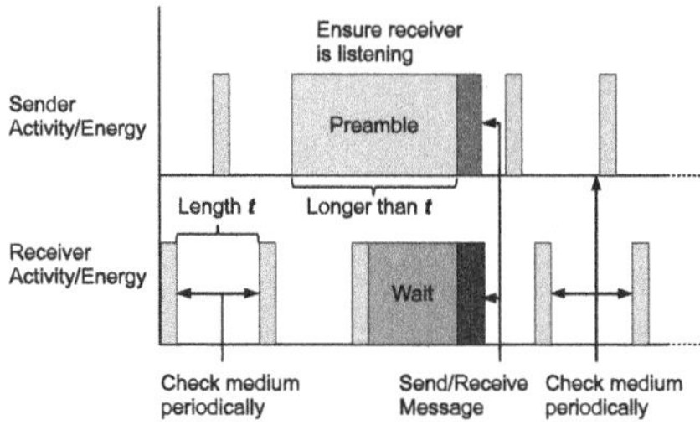


**Fig. 29** Illustration of B-MAC [46]

Since the radio chip is packet-based, it sends short packets to indicate that a.) no other node should send at this time and b.) the intended receivers (in this case all surrounding nodes since it is a broadcast) should keep the radio on. The length of the preamble is longer than 1 second to ensure that all receivers have time to switch their radio on. At $[T_3]$, the receiver wakes up the radio since 1 second has elapsed since the last wake up. Contrary to the last two times, this time the radio is kept on since the preamble of the sender indicates that the receiver is an intended recipient of a message. Broadcasting the actual message happens in a few milliseconds at $[T_4]$. Both nodes switch the radio off a few milliseconds after the message is broadcasted/received.

The readings from sender and receiver show two important points with regard to our hypotheses **H.2** and **H.3**: While the radio is switched on, both nodes consume an almost equal amount of energy, i.e., **H.2** is confirmed. Compared to the preamble of more than 1 second and the waiting for the actual message at the receiver, the time and energy spent for sending/receiving the message is negligible. An increase of the message size would result in a longer time spent sending the actual message. The time/energy spent previously for preamble and waiting is bigger by at least an order of magnitude unless hundreds of packets must be sent. With this we conclude that **H.3** is confirmed as well.

## A.3 Lessons Learned

This section investigated the energy consumption characteristics of sensor nodes in particular with regard to communication. As expected, communication has by far the largest impact on node lifetime. More accurately, the amount of time a sensor node has to switch the radio chip on significantly reduces its lifetime. As observed by [15] and our experiments, keeping the radio on at all times is not a viable option. The networking community has taken major steps to reduce idle listening, i.e., time where the radio chip is on but there no message to receive. While these efforts reduce idle listening and thereby increase node lifetimes significantly, they come at the cost of a large overhead for sending and receiving messages. Thus, the number of messages exchanged between sensor nodes is the most important factor for the evaluation of the energy efficiency of an application. Within reasonable limits, the actual size of the messages does not affect energy consumption significantly which is a common misconception, particularly of the database community.

This confirms our approach for the evaluation of our mechanisms for in-network processing: We count the number of messages sent and received to measure its impact on the sensor networks lifetime.